Samuel Newth

**A First Book of Natural Philosophy**

An introduction to the study of statics, dynamics, hydrostatics, optics, and

acoustics, with numerous examples

Samuel Newth

**A First Book of Natural Philosophy**
*An introduction to the study of statics, dynamics, hydrostatics, optics, and acoustics, with numerous examples*

ISBN/EAN: 9783337071264

Printed in Europe, USA, Canada, Australia, Japan

Cover: Foto ©berggeist007 / pixelio.de

More available books at **www.hansebooks.com**

# A FIRST BOOK

OF

# NATURAL PHILOSOPHY:

## AN INTRODUCTION

TO THE STUDY OF

## STATICS, DYNAMICS, HYDROSTATICS, LIGHT, HEAT, AND SOUND.

### With Numerous Examples.

BY

## SAMUEL NEWTH, M.A., D.D.,

FELLOW OF UNIVERSITY COLLEGE, LONDON ;
PRINCIPAL OF NEW COLLEGE, LONDON ;
ASSISTANT MATHEMATICAL EXAMINER IN THE UNIVERSITY OF LONDON.

New and Enlarged Edition.
(31st Thousand.)

LONDON:
JOHN MURRAY, ALBEMARLE STREET.

1879.

PLYMOUTH:
PRINTED BY WILLIAM BRENDON AND SON,
GEORGE STREET.

# PREFACE.

THE following work embraces all the subjects in Natural Philosophy required at the Matriculation examination of the University of London. For the present edition it has been revised throughout. Considerable additions have been made in nearly every chapter; and, in order to meet the more recent requirements of the University, a chapter on Heat has also been added. The demonstrations are carefully adapted to the requirements of those whose Mathematical knowledge does not extend beyond the First Book of Euclid, and the easier cases of Simple Equations. In a few instances, where Mathematical demonstrations present some perplexity to the beginner, experimental proofs have been substituted. Examples in illustration are given with all the more important propositions, and numerous other examples are added for exercise. I have endeavoured hereby to supply a work which shall also be generally useful as a First Book of Natural Philosophy. Some experience in teaching has confirmed the opinion, that the junior pupil may, with a twofold benefit, be early introduced to Natural Philosophy, as a branch of his Mathematical studies. The interesting applications which this study supplies greatly assist the beginner

in mastering the processes and results of pure Mathematics; and
his early familiarity with Mechanical principles, and with some
at least of their practical applications, enables him afterwards to
prosecute their more formal study with greater ease.

For the guidance of those who are preparing for University examinations I have introduced a considerable number of examples,
on all the subjects included in the volume, taken from recent
examination papers. These have been selected chiefly from
those set at the Matriculation examination; some are from the
Examinations for Women, and a few from the First B. Sc.
papers.

The chapter on Sound, though no longer required for Matriculation, is retained for the sake of those who may use the book
for other purposes.

SAMUEL NEWTH.

New College, London,
*October, 1877.*

# CONTENTS.

# APPENDIX.

# NATURAL PHILOSOPHY.

## CHAPTER I.

DEFINITIONS AND PRINCIPLES.

1. **Force.**—Whatever is capable of *producing* motion in a body, or any *change* in the motion of a body, is termed *force*.

In other words, force is the name we employ to express that unknown cause which, under any circumstances, can produce a change in the state, whether of rest or motion, of any material body.

Whatever causes a change in the motion of a body must be regarded as of like nature with that which produces motion, and hence the same term (force) is applied to both, even although there are some forces which, while they are able to change the motion of a body, can never produce it. Such, for example, are friction and resistances of all kinds. Forces of this nature can, it is evident, never act alone; for some other force must be present in order to produce the motion which they change, and hence if but one force act upon a body, it must be one capable of *producing* motion. Also, if one force only act upon a body, motion must necessarily follow.

2. **Equilibrium.**—When two or more forces act upon the same body, their united effect *may* be such that no motion ensues. Whenever this is the case, the forces are said to be in *equilibrium*.

3. **Statics and Dynamics.**—That branch of mechanics which investigates the relations which exist between forces in equilibrium is termed Statics; and that which investigates the

B

effects of forces not in equilibrium, but producing motion, is termed Dynamics.

**4. Pressure.**—Whenever motion is prevented by muscular effort, as, for instance, when a weight is held in the hand and so prevented from falling to the ground, or when an elastic cord is stretched and its rebound prevented, a sensation is produced which we call *pressure*. But, just as in several other cases, the same word is used both for the sensation and its cause—*e.g.*, sound, smell, taste—so is the term pressure applied also to the force whose resulting motion has been thus prevented. In this sense, then, pressure is force considered as the cause of the sensation which is felt when motion is prevented by muscular effort. It is with this meaning that the term is commonly employed in mechanics, although extended·to all forces when the motion they are capable of producing is in any way prevented, whether it be by muscular action or not. Thus, whether a weight be held in the hand, or rest upon a table, in either case it is said to exert a pressure, and a pressure of the same amount.

The commonest case of pressure is weight, and this supplies the most convenient standard of reference by which to compare different pressures. By means of weight, other pressures may in most cases be easily measured. Thus, when an elastic cord is held stretched by the hand, the pressure it exerts may be compared with others by determining what weight will keep the string stretched to the same degree.

Since all questions considered in Statics refer to forces in equilibrium, all statical forces may be denominated pressures, and consequently may be measured by weight.

**5. Representation of forces.**—Forces may differ from each other, not only in magnitude, but also in direction, and hence may be conveniently represented by straight lines; the direction of the lines representing the direction of the forces, and the lengths of the lines, each being measured by the same scale, the magnitudes of the forces.

Two forces are said to be *parallel*, when the lines which represent them are parallel; and are said to be *converging*, when the lines representing them are not parallel.

Two converging forces will be called *concurrent*, if they both push .towards or from the same point. Two parallel forces will be called concurrent, if they both push in the same direction.

6. **Tension.**—Force, when transmitted by means of a cord, is sometimes spoken of as the *tension* in the cord.

7. **Resultant.**—The single force which represents the combined effect of several forces is termed their *resultant;* relatively to the resultant, these several forces are termed *components* or *component forces. Composition of forces* takes place when two or more forces are replaced by a single force equivalent to them, that is, when the resultant is substituted for its components. *Resolution of forces* takes place when a single force is replaced by two or more forces equivalent to it, that is, when the components are substituted for their resultant.

8. **Rigid body.**—A body is called rigid, when the relative position of its particles is supposed to be unchangeable. In other words, a rigid body is one which is supposed to be perfectly hard, admitting of neither contraction, nor expansion, nor rupture, nor twisting.

9. PRINCIPLE I. *If two forces, equal in magnitude and opposite in direction, act upon a rigid body, they are in equilibrium : and conversely, if two forces, acting upon a rigid body, are in equilibrium, they are equal and opposite.*
This is the simplest possible case of equilibrium, and is assumed as an axiom.

10. It follows from the preceding, that if any number of forces be in equilibrium, any one of them must be equal in magnitude and opposite in direction to the resultant of the rest. For if, instead of the remaining forces, we substitute their resultant, we shall have but two forces; namely, this resultant and the force in question, and these are in equilibrium, and therefore are equal and opposite. The value of the resultant can hence be immediately deduced in certain simple cases. For example : *If two equal forces act at an angle of* 120°, *the resultant is equal to either of the components.*
Let OA, OB, and OC represent three equal forces acting upon O. Let the angles AOB, BOC be each 120°, then will AOC also be 120°. These three forces will be in equilibrium;

for there can be no reason why motion should take place in
one direction more than in another.  Hence, either one of the
forces, for instance OC, is equal and
opposite to the resultant of the other
two, OA and OB.  Therefore, OD, a
force equal and opposite to OC, is the
resultant of OA and OB.

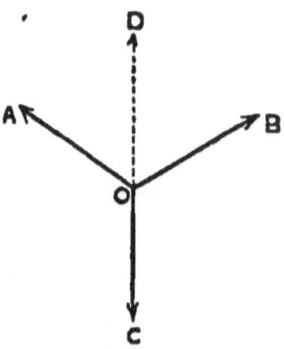

In a similar manner it may be
shown, that if four equal forces lying
in a plane act upon a point in such a
way that the angle between the first
and second, that between the second
and third, and that between the third
and fourth, are each equal to $72°$, the
resultant shall be equal to either of the components.

11. PRINCIPLE II.   *When forces are in equilibrium, the equilibrium will not be disturbed by the introduction or removal of
any number of forces that are themselves also in equilibrium.*

Forces in equilibrium can neither produce nor change motion ; and, consequently, the introduction of such a set of
forces will not cause motion ; nor, on the other hand, can
motion ensue from the removal of such a set of forces : for if it
could, that motion must have been checked by their presence,
or forces in equilibrium would have destroyed motion, which
they cannot do.

12. **Transmissibility of force.**—A force acting upon a
rigid body at rest may be supposed to act at any point within
the body in the line of its direction.  Thus, if a force of any
magnitude P act upon a rigid body at A, and along the line
AB in the direction of the arrow, we may suppose P to be
applied at any other point A′ in the line AB within the body.
For at A′ and B introduce two equal and opposite forces of the
same magnitude as P.  At A and B we have two equal and

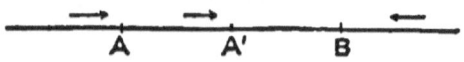

opposite forces.   These forces are (Principle I.) in equilibrium,
and may therefore (Principle II.) be removed.   There will then

remain only a force P at A' of the same magnitude as the original force, and acting in the same direction.

Also, if a force act at any given point of a perfectly flexible cord, it may be supposed to act at any other point in the cord taken on that side of the given point away from which the force pulls. Thus, if a weight of 1 lb. hang from any part of a flexible cord, the same effort is necessary to sustain it at whatever point above the weight the hand be applied. The same, also, would be the case, if the cord bearing the weight were passed over a perfectly smooth peg before it were held by the hand. That it is not so in an actual experiment arises from the absence of perfect flexibility in the cord, and of perfect smoothness in the peg.

**13. Greatest and least values of resultant.**—If two concurrent forces act along the same line, or along parallel lines, the resultant is their sum ; and if two non-concurrent forces act in the same line, or along parallel lines, the resultant is their difference. These are respectively the greatest and least values of the resultant of two forces.

When the two forces are converging, the resultant will have some value intermediate between these extremes. It will presently be seen that, as we should expect beforehand, the smaller the angle between the components, the more nearly will the resultant approach to the greatest possible value, and the larger the angle between the components, the more nearly will it approach the least possible value. The exact value of the resultant of two converging forces is found by aid of the principle commonly termed the *parallelogram of forces*.

**14.** PRINCIPLE III. **(Parallelogram of forces.)**—*If two concurrent forces acting upon a point are represented in magnitude and direction by the two sides of a parallelogram, then will their resultant also be represented in magnitude and direction by the diagonal drawn through the given point, and the resultant will be concurrent with the components.*

This proposition may be proved experimentally in the following manner. Let two cords, bearing the weights P and Q, pass over the fixed pulleys K and L, and be connected together at O. Let O, for convenience' sake, be temporarily fastened by a pin. Along OK mark off OA, representing the weight P on any scale ; that is, take as many divisions of the scale as there are pounds or ounces in P. Along OL mark off OB, representing Q on the same scale. Complete the parallelogram AOBC, and draw the diagonal CO. Let a third string be attached to O, and pass in the direction of CO produced over a pulley M, and a weight R be attached to it of the magnitude represented by the line CO. Then, if the pin be removed and O set at liberty, it will be always found that O remains at rest; that is, the forces P, Q, and R, acting upon it are

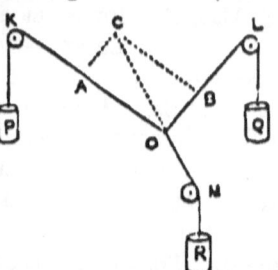

in equilibrium. But if three forces be in equilibrium, any one must be equal and opposite to the resultant of the other two. Consequently, R must be equal and opposite to the resultant of P and Q. Hence the diagonal CO represents the magnitude and direction of the resultant of the forces P and Q, represented respectively by the sides AO and BO.

### 15. Resultant of two converging forces determined graphically.

—By aid of the parallelogram of forces, the resultant of two given converging forces can in all cases be easily determined graphically. Produce the lines along which the forces act until they meet. From the point of intersection measure off distances along the two lines, such as shall represent on any scale the magnitudes of the two forces, taking care to measure these in the direction in which forces point when acting concurrently. Then complete the parallelogram of which these lines form two adjacent sides, and draw the diagonal through the above-mentioned point of intersection. Thus for example, let the two forces act in the lines OA and OB, intersecting in O. Let the one force act pointing from O to A, and the other pointing from O to B. Mark off the distances OC and OD representing the magnitudes of the two forces on any scale. Complete the parallelogram OCED. Then the diagonal OE represents the resultant upon the same scale. If the force in OB had pointed from B towards O, then the sides of the parallelogram

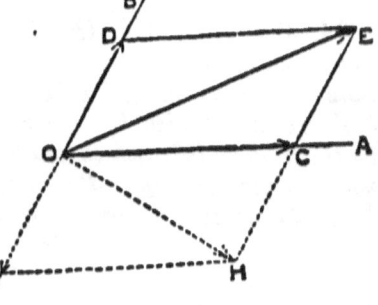

OCED would not have represented the forces in the way required by the parallelogram of forces. In this case we must have measured off the length which represents the force along BO produced, and completing the parallelogram OFHC, the resultant would be represented by the line OH.

### 16. Resultant of two converging forces determined arithmetically.

—The numerical value of the resultant of two given converging forces can, it has been seen, be found whenever the length of the diagonal of a parallelogram can be found. This in general requires a knowledge of trigonometry. There are, however, certain simple cases, for which a knowledge of the First Book of Euclid is sufficient. These are, when the angle between the two forces is one of the following angles, 90°, 30°, 150°, 60°, 120°, 45°, or 135°. The following examples will shew the method to be pursued.

*Ex.* 1. To find the resultant when forces of 48 and 36 lbs. act upon a point at right angles to each other.

Let OA and OB represent the two forces on any scale, then, completing the rectangle, OC will represent the resultant on the same scale. But OC is the hypotenuse of a right-angled triangle whose sides are 48 and 36; ∴ Euclid, B. i. Prop. 47,

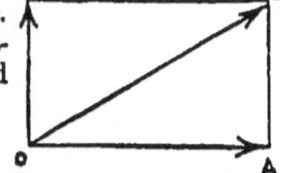

$$OC^2 = 36^2 + 48^2$$
$$\therefore \quad OC = \sqrt{(1296 + 2304)} = 60.$$

*Ex.* 2. To find the resultant when forces of 32 and 10 lbs. act upon a point at an angle of 60°.

Let OA = 32 and OB = 10; complete the parallelogram OBCA, and OC will represent the resultant.

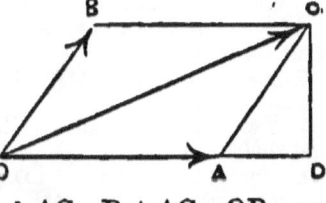

Draw CD at right angles to OA produced. Then, since the angle BOA is 60°, the angle CAD is also 60°, and the triangle CAD is half an equilateral triangle. Hence AD = ½ AC. But AC = OB = 10.
∴ AD = 5. Also $CD^2 = AC^2 - AD^2 = 100 - 25 = 75.$
But
$$OC^2 = OD^2 + CD^2$$
$$= 37^2 + 75 = 1444$$
$$\therefore \quad OC = 38.$$

*Ex.* 3. To find the resultant when forces of 20 and 12 lbs. act upon a point at an angle of 30°.

In the preceding figure, let $OA = 20$, $OB = 12$, and the angle $BOA = 30°$. Then $OC$ will represent the resultant. In the triangle $ADC$, the angle $CAD$ being equal to $BOA$ is 30°, and therefore the triangle is half an equilateral triangle, and therefore

$$CD = \tfrac{1}{2}AC = 6.$$

And

$$AD = \sqrt{(AC^2 - CD^2)}$$
$$= \sqrt{(144 - 36)} = 10 \cdot 392.$$

But

$$OC^2 = OD^2 + CD^2$$
$$= (30 \cdot 392)^2 + 36$$
$$= 959 \cdot 673664$$

∴

$$OC = 30 \cdot 978.$$

*Ex.* 4. To find the resultant when forces of 42 and 10 lbs. act upon a point at an angle of 45°.

As before, let $OA$ and $OB$ represent the two forces. Then the angle $CAD = 45°$, and ∴ $AD = CD$.

∴

$$CD^2 = \tfrac{1}{2} AC^2 = 50$$

And

$$AD = CD = 7 \cdot 071.$$

Therefore

$$OC^2 = (49 \cdot 071)^2 + 50$$
$$= 2457 \cdot 963041$$

∴

$$OC = 49 \cdot 577.^*$$

### EXAMPLES.

1. Find the resultant when forces of 60 and 144 lbs. act upon a point at right angles to each other.     *Ans.* 156 lbs.

2. Find the resultant when forces of 18 and 80 lbs. act upon a point at right angles to each other.     *Ans.* 82 lbs.

3. Two forces each of 100 lbs. act upon a point at an angle of 60°, find the resultant.     *Ans.* 173·2 lbs.

4. Forces of 36 and 60 lbs. respectively act upon a point at an angle of 60°, find the resultant.     *Ans.* 84 lbs.

5. Two forces each of 10 lbs. act upon a point at an angle of 45, find their resultant.     *Ans.* 18·477 lbs.

6. Two forces each of 30 lbs. act upon a point at an angle of 30°, find their resultant.     *Ans.* 57·955 lbs.

7. Two forces each of 100 lbs. act upon a point at an angle of 150°, find their resultant.     *Ans.* 51·7638 lbs.

8. Forces of 60 and 160 lbs. respectively act upon a point at an angle of 120°, find their resultant.     *Ans.* 140 lbs.

9. The resultant of two forces of 24 and 143 lbs. respectively is 145 lbs., shew that the angle between the forces is 90°.

* Examples 2, 3, and 4 may be solved more readily by aid of Euclid ii. 12, which gives

$$OC^2 = OA^2 + AC^2 + 2 \cdot OA \cdot AD.$$

10. The resultant of two forces of 35 and 165 lbs. respectively is 185 lbs., shew that the angle between the forces is 60°.

11. Two forces of 70 and 400 lbs. are kept in equilibrium by a force of 370 lbs., shew that the angle between the forces is 120°.*

### 17. Resolution of forces.

—Since the components may be substituted for their resultant, their effects being equivalent, it follows, from the parallelogram of forces, that for any force we may substitute two others, whose magnitude and direction are represented by the sides of any parallelogram, of which the line representing the given force forms the diagonal. And since an infinite number of parallelograms can be drawn, having a given line for their diagonal, any force can be resolved into two others, in an infinite number of ways.

Thus, if OE (fig. Art. 15) represent any force in magnitude and direction, and it be required to resolve it into two forces acting along the lines OA and OB, through E draw EC parallel to OB, and ED parallel to OA, then OC and OD will represent the two components required. It will be seen, that in a similar way a force can always be resolved into two others acting along any two lines drawn through any point in its own direction.

The following are examples of the resolution of forces arithmetically :

*Ex.* 1. A force of 200 lbs. is resolved into two others acting at right angles, one of the components is 56 lbs., required the other component.

Let OC (fig. Art. 16, Ex. 1) represent the given force, and OB the given component, then OA is the required component. But

$$OA^2 = OC^2 - AC^2$$
$$= (200)^2 - (56)^2$$
$$= 40000 - 3136 = 36864$$
$$\therefore \quad OA = 192.$$

*Ex.* 2. A force of 100 lbs. is resolved into two others at right angles, one of the components is inclined to the given force at an angle of 30°, find the components.

---

* It will aid the student in the solution of these and the following examples to remember that if any right angled triangle have its other angles 30° and 60°, the side opposite 30° is equal to half the hypotenuse, and the side opposite 60° is equal to half the hypotenuse multiplied by $\sqrt{3}$. In any right angled triangle, having its angles 45°, each side is equal to half the hypotenuse multiplied by $\sqrt{2}$.

Let OC in the same figure be the given force, and let the angle COA be 30°, then OA and OB are the required components.

Since COA is 30° and CAO is a right angle, the triangle OAC is half an equilateral triangle, and $AC = \frac{1}{2}OC$. But $OB = AC$. Therefore,

$$OB = \frac{1}{2}OC = 50$$

And
$$OA = \sqrt{(OC^2 - AC^2)} = \sqrt{7500}$$
$$= 86\cdot6.$$

*Ex.* 3. A force of 76 lbs. is resolved into two others acting at an angle of 60°, one of the components is 20 lbs., required the other component.

Let OC (fig. Art. 16, Ex. 2) represent the given force, and OB the given component. Let the angle BOA = 60°; then OA will represent the required component. To find OA, draw CD at right angles to OA produced. Then, since the angle CAD is 60°, the triangle CAD is half an equilateral triangle, and $AD = \frac{1}{2}AC$. But $AC = OB = 20$ $\therefore$ $AD = 10$, and

$$CD^2 = AC^2 - AD^2$$
$$= (20)^2 - (10)^2 = 300.$$

Also,
$$OD^2 = OC^2 - CD^2$$
$$= (76)^2 - 300 = 5476$$

$\therefore$
$$OD = 74$$

But
$$OA = OD - AD$$
$$= 74 - 10 = 64.$$

## EXAMPLES.

1. If a force of 100 lbs. be resolved into two equal forces at right angles, what is the magnitude of either component?
Ans. 70·71 lbs.

2. If a force of 100 lbs. be resolved into two equal forces acting at an angle of 60°, what is the magnitude of either component?
Ans. 57·735 lbs.

3. If a force of 100 lbs. be resolved into two equal forces acting at an angle of 30°, what is the magnitude of either component?
Ans. 51·76 lbs.

4. If a force of 100 lbs. be resolved into two equal forces acting at an angle of 135°, what is the magnitude of either component?
Ans. 130·659 lbs.

5. The resultant of two forces at right angles is 145 lbs., one of the components is 24 lbs., find the other component. Ans. 143 lbs.

6. The resultant of two forces acting at an angle of 60° is 42 lbs., one of the components is 18 lbs., find the other component. Ans. 30 lbs.

7. The resultant of two forces acting at an angle of 120° is 84 lbs., one of the components is 36 lbs., find the other component.

Ans. 96 lbs.

8. The resultant of two forces acting at an angle of 30° is 70 lbs., one of the components is 20 lbs., find the other component.  Ans. 51·96 lbs.

9. The resultant of two forces acting at an angle of 150° is 70 lbs., one of the components is 20 lbs., find the other component.

Ans. 86·6 lbs.

10. The resultant of two forces acting at right angles is 20 lbs., and makes an angle of 30° with one of the components, find the two components.  Ans. 10 and 17·32 lbs.

**18. Composition of three or more forces acting upon a point.**—If three forces act upon a point, and their resultant be required, find the resultant of any two of them; the composition of this resultant with the third force will give the resultant of the three given forces.

Thus, let OA, OB, OC be the three forces.   First, complete the parallelogram OADB, then OD will represent the resultant of the two forces OA and OB. Secondly, complete the parallelogram ODEC, then OE will represent the resultant of OD and OC; that is to say, the resultant of the three forces OA, OB, and OC.

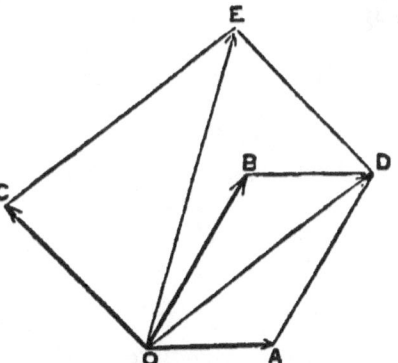

In a similar way, the resultant of any number of forces acting upon a point may be determined graphically.    Another, and sometimes more convenient, method is given in Art. 23.

19. *Def.* The sides of any rectilinear figure are said to be taken *in order*, when taken as they would be traversed by a point moving continuously around the figure in either direction; that is to say, either as the hands of a watch revolve, or in the contrary direction. Thus, if ABCD be any quadrilateral, its sides taken in order are either AB, BC, CD, DA, or AD, DC, CB, BA. It will be sometimes convenient to describe these relatively to each other, as in direct and reverse order respectively.

20. *If two sides of a triangle, taken in order, represent in magnitude and direction two forces acting upon a point, then*

*shall the third side, taken in reverse order, represent the resultant in magnitude and direction.*

Let the sides KL, LM, of the triangle KLM, represent in magnitude and direction the forces P and Q acting at any point O, then will KM represent in magnitude and direction the resultant of P and Q. Through O draw OA equal and parallel to KL, and pointing in the same way; also OB equal and parallel to LM, and pointing in the same way: then OA and OB represent P and Q respectively. Completing the parallelogram, the diagonal OC represents the resultant of P and Q. But since the lines OA, AC, are parallel respectively to KL, LM, and point in the same way, the angle OAC is equal to the angle KLM. Then in the triangles KLM, OAC, the sides KL, LM, and the angle KLM are equal severally to the sides OA, AC, and the angle OAC; therefore KM is equal to OC. It can also be easily shewn that KM is parallel to OC; therefore the line KM represents OC, the resultant of P and Q, both in magnitude and direction.

21. PRINCIPLE IV. **Triangle of forces.**—*If three forces, represented in magnitude and direction by the sides of a triangle taken in order, act upon a point, they will be in equilibrium ; and, conversely, if three forces acting upon a point, and in equilibrium, be represented in direction by the sides of a triangle taken in order, they will also be represented in magnitude by the sides of that triangle.*

*First.* Let three forces, represented in magnitude and direction by the sides of the triangle ABC, taken in order, that is to say, by AB, BC, and CA, act upon any point, these forces will be in equilibrium.

For, by the preceding article, the resultant of the forces represented by AB and BC is a force represented by AC; that is, a force equal and opposite to the force represented by CA. The forces are therefore in equilibrium.

*Secondly.* Let any three forces, P, Q, and R, acting upon a point, be in equilibrium, and let the sides of the triangle ABC, taken in order, represent the direction of these forces, viz., AB the direction of P; BC, that of Q; and CA, that of R. Then will these sides, AB, BC, CA, represent severally the magnitudes of P, Q, and R.

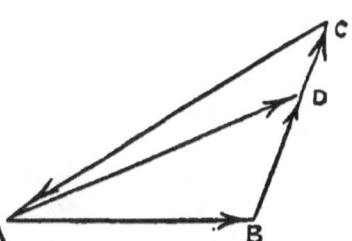

Let AB represent P on any scale, then must BC represent Q on the same scale. For, if possible, let some other length BD represent Q; then, since AB, BD, two sides of a triangle taken in order, represent in magnitude and direction two forces acting upon a point, the third side AD, by Art. 20, represents their resultant. But, by hypothesis, there is equilibrium; and therefore a force represented in direction by AD is balanced by a force represented in direction by CA, which is impossible. Therefore, BD cannot represent the magnitude of Q. In like manner it may be shewn, that no other length than BC can represent Q. Then, since AB represents P, and BC represents Q, it follows that CA represents R.

If a triangle be drawn with one of its sides equal to AB, and its sides severally perpendicular to the sides of the triangle ABC, this triangle will be equal to ABC in all respects. Hence, *if three forces be in equilibrium, and any triangle be drawn whose sides are severally either parallel or perpendicular to their directions, the forces are to one another as the sides of the triangle.*

The "triangle of forces" is but the "parallelogram of forces" in another form, and is generally the more convenient form for the solution of problems in which three forces acting in known directions are in equilibrium, and the magnitude of one of them being given, the magnitude of the others is required. The following are examples in illustration:

*Ex.* 1. A weight of 100 lbs. is sustained by two cords, whose lengths are severally 21 and 28 inches, fastened to two points lying in the same horizontal line, required the tensions in the cords when the cords are at right angles.

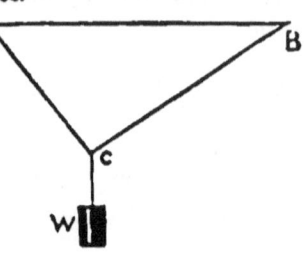

Let CA and CB be the two cords, the line AB being horizontal, and the angle ACB a right angle. Let W be the weight of 100 lbs. The point C is at rest under the action of three forces; viz., W acting vertically, and the tensions in CA and CB. In the triangle ACB

       AB is perpendicular to W,
       BC        ,,        the tension in CA,
       CA        ,,        the tension in CB;

therefore, the three forces are represented on the same scale by the sides of this triangle. Hence

$$\text{tension in CA} : 100 : : BC : AB.$$

But BC = 28, AC = 21, and the angle ACB is a right angle, therefore $AB^2 = (21)^2 + (28)^2$; whence AB = 35.

$\therefore$        tension in CA : 100 : : 28 : 35,
$\therefore$        tension in CA = 80 lbs.

Similarly,

       tension in CB : 100 : : CA : AB
                        : : 21 : 35
$\therefore$        tension in CB = 60 lbs.

*Ex.* 2. A cord having equal weights attached to its extremities passes over pulleys placed at A and B in the same horizontal line, 70 inches apart, and through a smooth ring C, from which a weight of 100 lbs. is attached; what must be the magnitude of the equal weights, that C may rest exactly 12 inches below the line AB?

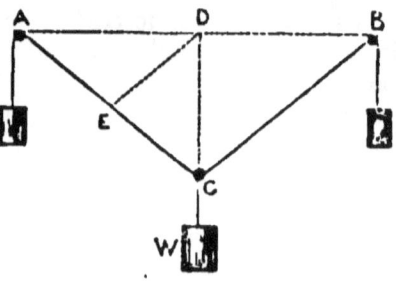

Let W be the weight of 100 lbs. Let the vertical line through C meet the line AB in D; then, since there is no reason why C should be nearer to A than to B, AB will be bisected in D. Through D draw DE parallel to BC, then AC is bisected in E.

In the triangle CDE
DC is parallel to W,
CE     „       tension in CA,
ED     „       tension in CB;
therefore,

tension in CA : W : : CE : CD.
But by hypothesis $CD = 12$. Also $CE = \frac{1}{2}AC$, and $AC = \sqrt{(AD^2 + CD^2)} = \sqrt{(1225 + 144)} = 37.$
Hence

tension in CA : 100 : : $18\frac{1}{2}$ : 12 ;
∴     tension in CA = $154\frac{1}{6}$ lbs.

But (Art. 12) the tension in the cord is equal to the weight hanging from its extremity, and therefore the required weight is $154\frac{1}{6}$ lbs.

*Ex.* 3. A ball weighing 20 lbs. slides along a perfectly smooth rod inclined at an angle of 30° with the vertical line, what force applied in the direction of the rod will sustain the ball, and what is the pressure upon the rod?

Let A be the ball, sustained on the rod by the force P. Then the ball is at rest under the action of three forces; namely, its weight acting vertically downwards, the resistance of the rod acting at right angles to the rod, and P acting along the rod.

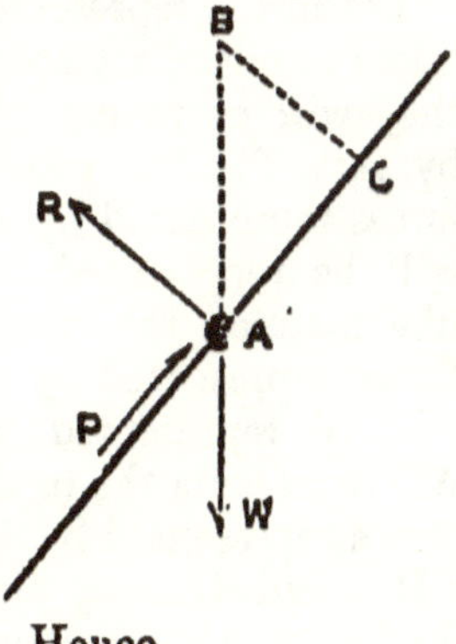

In the vertical line through A take any point B, and draw BC perpendicular to the rod. Then in the triangle ABC
AB is parallel to W,
BC     „     R,
CA     „     P ;
therefore, P is represented by CA, and R by BC, on the same scale that AB represents W or 20. But if $AB = 20$, then, because the angle BAC is 30°, $BC = \frac{1}{2}AB = 10$, and $CA = \sqrt{300} = 17.32$. Hence,

$P = 17\cdot32$ and $R = 10.$

## EXAMPLES.

1. If a weight of 40 lbs. slide along a perfectly smooth rod, inclined at an angle of 60° with the vertical line, what force is necessary to sustain the weight, and what is the pressure on the rod?

Ans. 20 and 34·64 lbs.

2. In the previous question, if the rod were inclined at an angle of 45°, shew that the force necessary to sustain the weight and the pressure on the rod will each be 28·284 lbs.

3. A smooth ring sustaining a weight of 50 lbs. slides along a cord fastened at two points, lying in the same horizontal line, find the tension in the cord when the two parts of the cord are at right angles to each other.      Ans 35·35 lbs.

4. In the previous question, what will be the tension in the cord when the two parts of the cord form an angle of 60°?  Ans. 28·9 lbs. nearly.

5. One extremity of two cords, whose lengths are severally 14 and 48 inches, is fastened at points in the same horizontal line ; the other extremity is fastened to a weight of 100 lbs., and the cords are at right angles, find the tensions in the cords.

Ans. 96 lbs. in the shorter cord, and 28 in the longer.

6. Find the horizontal and vertical pressures, when a force of 100 lbs. acts in a direction making an angle of 60° with the vertical line.

Ans. 86·6 and 50 lbs.

7. Find the horizontal and vertical pressures, when a force of 60 lbs. acts in a direction making an angle of 45° with the vertical line.

Ans. each equals 42·426 lbs.

8. Find the horizontal and vertical pressures, when a force of 80 lbs. acts in a direction making an angle of 30° with the vertical line.

Ans. 40 and 69·28 lbs.

**22. Polygon of forces.**—*If any number of forces acting upon a point be represented in magnitude and direction by the sides of a polygon taken in order, they will be in equilibrium.*

Let forces represented by the sides of the polygon ABCDE, taken in order, act upon any point, they will be in equilibrium ; for, by Art. 20, the resultant of the forces represented by AB and BC will be represented by AC. In like manner, the resultant of the forces represented by AC and CD will be represented by AD ; or AD represents the resultant of the forces represented by AB, BC, and CD. Substituting this resultant

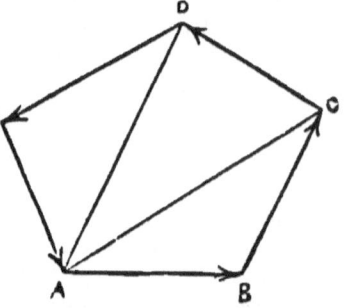

for its components, we have remaining three forces, represented by AD, DE, EA, three sides of a triangle taken in order ; and, therefore, by the triangle of forces, they are in equilibrium.

23. If any number of forces be in equilibrium, a force equal and opposite to any one will be the resultant of the remaining forces. Hence any side of a polygon, taken in reverse order, will represent the magnitude and direction of the resultant of

any number of forces acting upon a point, when these forces are represented in magnitude and direction by the remaining sides of the polygon taken in direct order.

Thus, if AB, BC, CD, and DE in the last figure represent in magnitude and direction four forces acting upon any point, the remaining side AE (not EA) will represent the magnitude and direction of their resultant.

Hence, to find the magnitude and direction of any number of forces acting concurrently upon any point, draw a line parallel to one of the forces, and representing it in magnitude. Through that extremity of this line which points in the same way as the force, draw a second line parallel to the next force, pointing in the same way, and representing it in magnitude. Through the extremity of this second line draw a third line parallel to the third force, pointing in the same way, and also representing it in magnitude. Proceed in this way until lines have been drawn representing all the forces. The straight line which completes the polygon, taken in reverse order, will represent the resultant in magnitude and direction.

**24. Resultant of three or more forces acting upon a point determined arithmetically.**—When the arithmetical value of the resultant of any number of forces acting upon a given point is required, the most convenient method is to resolve each force into two others acting along two lines (or axes) at right angles, and passing through the given point. The resultant of each of these sets of forces is readily found, being their algebraic sum; and the given forces, however many in number, are thus reduced to two forces at right angles. The resultant of these, as seen by *Ex.* 1, Art. 16, is the square root of the sum of the squares of the two forces.

Usually it will be most convenient to take the direction of one of the forces for one of the axes, and a line at right angles to this for the other. If two of the given forces be at right angles, their directions will be the most convenient axes.

### EXAMPLES.

1. Three forces, each of 10 lbs., act upon a point, the angle between the first and second is 30°, and the angle between the second and third is 60°, find the resultant. Ans. 23·9 lbs.

2. Three forces, each of 10 lbs., act upon a point, the angle between the first and second is 90°, and the angle between the second and third is 30°, find the resultant. Ans. 19·3 lbs.

C

3. If in the last example the angle between the second and third force be 60°, what is the resultant?                                              Ans. 15·03 lbs.

4. Forces 4, 2, and 1 act upon a point, the angle between the first and second, and the angle between the second and third, are each 60°, what is the resultant?                                                        Ans. 3√3.

5. Four forces, each of 20 lbs., act upon a point, the angle between the first and second is 30', the angle between the second and third is 30°, and the angle between the third and fourth is 60°, find the resultant.
                                                        Ans. 58·2 lbs. nearly.

6. Four forces, 20, 20, 10, and 10, act upon a point, the angle between the first and second is 45°, the angle between the second and third is 75°, and the angle between the third and fourth is 120°, find the resultant.
                                                        Ans. 28 nearly.

**25. Moment of a force about a point.**—The product of a force, and the perpendicular distance of its direction from any given point, is termed the *moment* of the force about that point. For instance, if a force exert a pressure of 20 lbs., and the perpendicular drawn to its direction from any point be 6 inches, the moment of the force about that point is 20 × 6, or 120. The moment of a force about a point measures the effect of the force in turning the body round that point, so that if the moments of two forces about any point be equal, they will each exert the same effort in turning the body around the point.

As the resultant of any number of forces acting upon a rigid body produces the same effect as the components, it follows, from what has been just stated, that (Principle V.) *the moment of the resultant about any point is equal to the sum of the moments of the components about that point.*

In the application of this principle, attention must be given to the direction in which each force tends to produce rotation. When all the forces are supposed to act in the same plane, it is only possible that the body can rotate in one or other of two directions, and these may therefore be distinguished as positive and negative. It is common to take the direction in which the hands of a watch rotate as the negative direction, and the contrary as the positive direction.

The moment of a force about any point in its own direction is of course zero, and hence *the sum of the moments of any number of forces about a point in the direction of their resultant is zero.* When the components are only two in number, in order that the sum of their moments may be zero, these moments must be of contrary sign and of equal value, and hence *the moments of two forces about any point in the direction of their resultant are equal : and conversely, if the moments of two forces about any point in their plane be equal and of contrary sign, that point will lie in the direction of their resultant.*

This may be deduced from the parallelogram of forces in the following way.*

Let P and Q be two forces acting respectively in the lines AO and BO, and let CO be the direction of their resultant. In CO take any point D, and from D draw the perpendiculars DG and DH. Let $DG = p$, and $DH = q$.

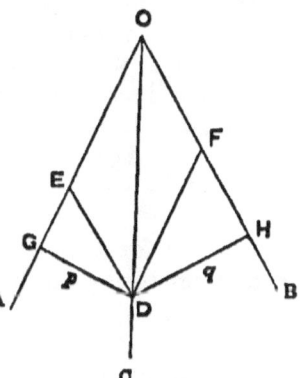

Then shall $Pp = Qq$.

For draw DE parallel to OB, and DF parallel to OA. Then, since the resultant of P and Q passes along the diagonal OD, the sides of the parallelogram OEDF will represent the forces. Therefore OE = P, and OF = Q. The triangles OED, OFD are equal. But the triangle OED = $\frac{1}{2}$ OE × $p$,† and the triangle OFD = $\frac{1}{2}$ OF × $q$.

$$\therefore \quad \tfrac{1}{2} OE \times p = \tfrac{1}{2} OF \times q.$$
$$\therefore \quad OE \times p = OF \times q.$$
$$\therefore \quad Pp = Qq.$$

Conversely. If $Pp = Qq$, the resultant of P and Q passes through D.

For $OE \times p = OF \times q$,

And $Pp = Qq$,

$$\therefore \quad \frac{OE}{P} = \frac{OF}{Q}$$

$$OE : OF :: P : Q;$$

that is, if OE represent P on any scale, OF will represent Q on the same scale; and, consequently, by the parallelogram of forces, their resultant will pass along the diagonal OD.

26. If a body turn about a fixed point or pivot, it is clear that no force, whose direction passes through the fixed point, will cause any motion. Consequently, if two or more forces act upon such a body, and their resultant pass through the fixed point, no motion will be caused. But, by Principle V., the resultant will pass through the fixed point if the sum of the

* The student may omit this on a first reading.
† The area of any triangle equals half the product of the base and height.

moments of the forces about that point is zero. This, therefore, is the only condition of equilibrium in such a case.

*Ex.* Forces of 20, 12, and 10 lbs. act upon a rigid body at
distances of 14, 15, and 16 inches respectively from a certain
fixed point in the body; the first and second tend to produce
revolution in the positive direction; and the third in the negative
direction; at what distance must a force of 15 lbs. act, in order
that the body may remain at rest?

Let $x$ be the required distance. Then, by Principle V.,

$$15x + 20 \times 14 + 12 \times 15 - 10 \times 16 = 0,$$
$$\therefore \qquad 15x = -300,$$
$$x = -20;$$

that is, the force must act at a distance of 20 inches, and must
tend to produce rotation in the negative direction.

### EXAMPLES.

1. Two forces, P and Q, act upon a body which moves about a fixed
point, at distances of 8 and 10 inches respectively, what must Q be, if P
is 50 lbs. when there is equilibrium?　　　　　　　Ans. 40 lbs.

2. A force of 80 lbs. acts at a distance of 12 inches from a fixed point,
at what distance must a force of 60 lbs. act, in order that there may be
equilibrium?　　　　　　　　　　　　　　　　　Ans. 16 in.

3. Forces of 18 and 26 lbs. respectively, acting at right angles to the
arms of a bent bar moving about the vertex, are in equilibrium, the
shorter arm is 9 in., what is the length of the longer arm?
　　　　　　　　　　　　　　　　　　　　　　Ans. 13 in.

4. Forces of 63 and 35 lbs. acting at the extremities of a straight bar,
at angles of 45° with the bar, are in equilibrium, the bar moves about a
pivot placed 5 in. from one extremity, what is the length of the bar?
　　　　　　　　　　　　　　　　　　　　　　Ans. 14 in.

5. A straight bar, 13 inches in length, moves about a point placed 3
inches from one extremity, a force of 30 lbs. acts at the end of the longer
arm, but inclined to it at an angle of 45°, what force acting at the shorter
arm, at an angle of 30°, will preserve equilibrium?　　Ans. 141·42 lbs.

6. A circular plate, whose radius is 20 inches, is moveable about its
centre C in a vertical plane; from two points A and B in the circumference, 90° apart, weights of 18 and 24 lbs. respectively are suspended
by cords, find the distance of the lighter weight above or below the
horizontal line through C when the plate is at rest.　　Ans. 12 in.

### 27. Composition and resolution of parallel forces.—
The resultant of two concurrent parallel forces is, it has already
been stated, their sum; and of two non-concurrent parallel forces,
their difference. It hence follows that the resultant of any
number of concurrent parallel forces is their sum also; and that

the resultant of any number of parallel forces, some of which act in one direction, and others in the contrary direction, may be found by taking the sum of those which act in the one direction, and subtracting the sum of those which act in the other direction.

Having in this way found the magnitude of the resultant, its position may be determined by means of the principle that the moment of the resultant about any point is equal to the sum of the moments of the components. (Art. 25.)

Thus, if R, $P_1$, and $P_2$, be any three parallel forces acting at the points C, A, and B in the straight line CO drawn at right angles to their directions :—

Then the magnitude of the resultant is

$$R + P_1 - P_2 ;$$

and if G be the point in which the resultant meets the line CO, the position of G may be found by taking the moments about any point whatever in the line CO, say the point O. Then

$$(R + P_1 - P_2) \times GO = R \times CO + P_1 \times AO - P_2 \times BO ;$$
$$\therefore \quad GO = \frac{R \times CO + P_1 \times AO - P_2 \times BO}{R + P_1 - P_2}.$$

N.B.—It will generally be found to be most convenient to take the moments about a point in one of the forces (in the preceding figure either C or B), in which case the moment of this force will be zero. Thus, to find the distance of G from B,

$$(R + P_1 - P_2) \times GB = R \times CB + P_1 \times AB ;$$
$$\therefore \quad GB = \frac{R \times CB + P_1 \times AB}{R + P_1 - P_2}.$$

*Ex.* Weights of 6, 8, and 10 lbs. respectively are suspended from the points A, B, and C, in the same horizontal line, required the distance of the resultant from A, AB and BC being each 12 inches.

The magnitude of the resultant is $6 + 8 + 10 = 24$; and if $x$ be the required distance,

$$24x = 6 \times 0 + 8 \times 12 + 10 \times 24$$
$$= 336 ;$$
$$\therefore \quad x = 14.$$

28. The resultant of two concurrent parallel forces passes between ·them, for (Art. 25) the moments of two components about any point in their resultant are equal and of contrary sign, and this can only be when the point lies between them; and for a similar reason the resultant of two non-concurrent parallel forces passes outside the greater.

The preceding results, respecting the magnitude and position of the resultant of two parallel forces, may be deduced from the parallelogram of forces in the following way.*

Let P and Q be any two concurrent parallel forces represented by the lines GA and HB. Draw any line AB at right angles to their directions. At A and B introduce two equal and opposite forces (Principle II.) represented by 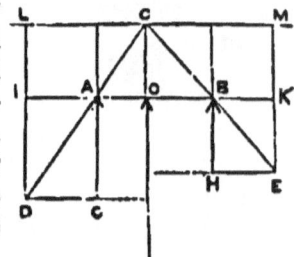 IA and KB. The resultant of IA and GA is DA, and the resultant of KB and HB is EB. Remove the points of application of these two resultants from A and B to C. Resolve these into their original components, and we shall have acting along LM two equal and opposite forces, corresponding to IA and KB, which, being in equilibrium, may be removed, and two forces equal respectively to P and Q acting along CO. The resultant, therefore, of P and Q is a force of the magnitude P + Q acting along CO.

The parallelograms GO and AL are equal, being complements of the parallelograms about the diagonal DC. Similarly, the parallelograms HO and BM are equal. But the parallelograms AL and BM are equal, being upon equal bases and between the same parallels; therefore, the parallelograms GO and HO are equal. But GO is a rectangle, and therefore equals GA × AO; and similarly HO = HB × BO. Hence,

$$GA \times AO = HB \times BO;$$
$$\text{that is, } P \times AO = Q \times BO,$$

or the moments of the components about a point O in the direction of the resultant are equal.

In the preceding, let R stand for the resultant of P and Q. Then, if R act at O, and if P and Q act at A and B in a contrary direction to that represented in the figure, P, Q, and R will

* This may be omitted on the first reading.

be in equilibrium. But when three forces are in equilibrium, any one is equal and opposite to the resultant of the other two. Consequently, GA, a force concurrent with R, will represent the resultant of the two non-concurrent forces R and Q. But P equals GA, therefore the resultant of R and Q equals P; but R equals P + Q, therefore P equals R − Q, or the resultant of the two non-concurrent forces R and Q is equal to their difference. As already shown, P × AO = Q × BO; add to both Q × AO, then

$$P \times AO + Q \times AO = Q \times BO + Q \times AO$$
$$(P + Q) \, AO = Q \, (BO + AO)$$
$$R \times AO = Q \times AB,$$

or the moments of the components about a point A in their resultant are equal.

29. *Two parallel forces*, P *and* Q, *act at a distance* a *from each other, to find the distance of the resultant from either force.*

Let R be the resultant, and $x$ its distance from any point in P; then, taking the moments about this point, therefore

$$Rx = Qa$$
$$x = \frac{Qa}{R};$$

or the distance of the resultant from one of the components equals the distance between the components multiplied by the other component and divided by the resultant.

30. *To resolve a force into two parallel forces acting at given points.*

Let R be the given force acting at the point C, and let A and B be the given points. Let P and Q be the required forces acting at A and B respectively. Then, taking the moments about B,

$$R \times CB = P \times AB \therefore P = \frac{R \times CB}{AB}.$$

Similarly, taking the moments about A,

$$R \times CA = Q \times BA \therefore Q = \frac{R \times CA}{BA}.$$

Hence, to find either component, multiply the given force by the distance of the other component, and divide by the distance between the two components.

When A and B are on opposite sides of C, P and Q are both

concurrent with R; but when A and B are on the same side of C, that force only which is the nearer to R will be concurrent with it, the other will be non-concurrent.

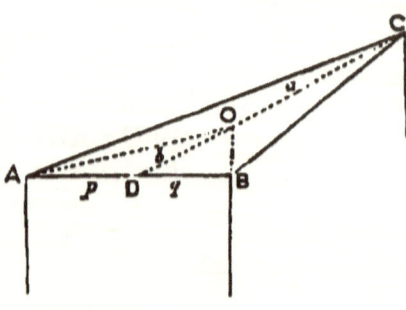

*Ex.* A weight of 1000 lbs. is placed upon the triangular table ABC at O, to find the pressures upon the three legs, when $a$ is 30 inches, $b$ 10 inches, $p$ 13 inches, and $q$ 12 inches.

Resolving 1000 lbs. at O into pressures at C and D, we obtain

$$\text{pressure at } C = \frac{1000 \times 10}{40} = 250,$$

$$\text{pressure at } D = \frac{1000 \times 30}{40} = 750.$$

Similarly, resolving the pressure at D into pressures at A and B

$$\text{pressure at } B = \frac{750 \times 13}{25} = 390,$$

$$\text{pressure at } A = \frac{750 \times 12}{25} = 360.$$

## EXAMPLES.

1. Find the position of the resultant of two concurrent parallel forces whose magnitudes are 245 and 125 lbs. respectively, acting at points 74 inches apart.　　　　　Ans. 25 in. from the larger force.

2. Two non-concurrent parallel forces, whose magnitudes are 340 and 120 lbs. respectively, act at points 44 inches apart; find the position of the resultant.　　　　Ans. 24 in. from the larger force.

3. The larger of two concurrent parallel forces is a pressure of 28 lbs., the resultant is equal to 40 lbs. acting at a distance of 18 inches from the larger component; what is the distance between the components?
　　　　　　　　　　　　　　Ans. 60 in.

4. The larger of two non-concurrent parallel forces is 70, the resultant is 49, acting at a distance 15 from the larger component; what is the distance between the components?　　　　Ans. 35.

5. The smaller of two non-concurrent parallel forces is 15, the resultant is 45, acting at a distance 56 from the smaller component; what is the distance between the components?　　　　Ans. 42.

6. Resolve a force of 700 lbs. into two concurrent parallel forces acting at distances from it of 17 and 3.　　　　Ans. 595 and 105 lbs.

7. Resolve a force of 1,000 lbs. into two non-concurrent parallel forces acting at distances from it of 4 and 24.　　Ans. 1,200 and 200 lbs.

8. A weight of 648 lbs. is placed upon the triangular table ABC (*fig. art.* 30) at O ; what are the pressures on the legs, when $a=27$ inches, $b=9$, $p=10$, and $q=17$?
Ans. pressure on A $=306$ ; pressure on B $=180$ ; pressure on C $=162$ lbs.

9. A beam AB, 10 feet in length, rests horizontally upon two vertical props, A and B, and another beam, CD, 20 feet in length, rests upon two vertical props, C and D ; a third beam, 30 feet in length, lies across the two former beams in such a way that the one extremity E is 3 feet from the prop A, and the other extremity F is 5 feet from the prop D ; a weight of 60 lbs. is attached to the third beam at a distance of 10 feet from F : shew that the pressures upon A, B, C, D are 14, 6, 10, and 30 lbs. respectively.

10. A weight resting upon a horizontal table exerts an equal pressure upon each of the three legs ; determine the position of the weight.

11. If P and Q be concurrent parallel forces acting at distances of 8 in. and 20 in. respectively from a point lying without them, what is the distance of the resultant from the same point when $P=20$ and $Q=30$?
Ans. 15·2 in.

12. If P, Q, and R be concurrent parallel forces, acting at distances of 6 in., 8 in., and 10 in. respectively from a point lying without them, what is the distance of the resultant from the same point when $P=12$, $Q=10$, and $R=18$?
Ans. 8·3 in.

13. A weightless beam, 12 feet in length, is suspended in a horizontal position by two cords attached at a distance of 1 foot from each end, weights of 10 and 20 lbs. are suspended from the ends of the beam, what are the tensions in the cords?
Ans. 9 lbs. and 21 lbs.

14. A beam AB, 12 feet in length, and weighing 6 lbs., rests horizontally upon two props, one placed 2 feet from A, and the other 3 feet from B, what are the pressures upon the props when a weight of 10 lbs. is suspended from A, and a weight of 20 lbs. from B, the weight of the beam itself being supposed to act at its middle point?
Ans. 6⅞ lbs. on prop near to A and 29⅛ on prop near to B.

# CHAPTER II.

ON THE CENTRE OF GRAVITY.

**31. Centre of parallel forces.**—If any number of parallel forces act at given points in a rigid body, it can be shewn that there is a point through which the resultant will always pass, whatever the position in which the body may be placed. This point is termed the centre of the parallel forces.

**32. Gravity and weight.**—*Gravity* is the force with which every particle of matter is drawn towards the earth. The *weight* of a body is the total force with which that body is drawn towards the earth : it is therefore the same as the resultant of the forces which, in consequence of the existence of gravity, act upon its several particles.

The force of gravity is not a constant force ; it is diminished by an increase in the distance from the earth, and it is increased by an increase in the latitude of the place of observation. Hence also the weight of a body is not constant ; the same body weighs less on the top of a mountain than at its base ; and, other things being equal, weighs more in London than in Jamaica. A body which at the equator and at the sea level weighs 294 lbs. will weigh at the same level in the latitude of London 295 lbs. Part of this increase is due to the fact that a point on the sea level at London is nearer to the centre of the earth than a point on the sea level at the equator. A body weighing 500 lbs. at the equator will in London, and at the same distance from the centre of the earth, weigh rather more than 501 lbs. This variation in the weight of a body will not of course be manifest when a body is weighed by an ordinary pair of scales, since the weights will be as much affected as the body to be weighed. To detect it, some other means, as for instance a spring balance, must be employed.

**33. Density.**—If two bodies of the same size be unequal in

weight, that one which is the heavier is said to have the greater
density. A body is said to have a uniform density if the weight
of any assignable portion whatever, say a cubic inch, is the
same in all parts of the body. If equal portions have different
weights in different parts of the body, the density is said to be
variable.

34. **Centre of gravity.**—The direction in which gravity
acts at any point is called the vertical line through that point.
This is nearly, though not strictly, the same as the line drawn
from the point to the earth's centre, and hence it is commonly
said that gravity acts towards the centre of the earth. The
forces, therefore, which in consequence of gravity act on the
several particles of any body are not, strictly speaking, parallel
forces, but, when the body is of moderate dimensions, they are
so nearly parallel that they may be regarded as such without
any sensible error. The centre of these parallel forces is what
is termed the *centre of gravity.*

The centre of gravity of a body is, then, the point through
which, in every position of the body, will pass the resultant of
all the forces which, in consequence of gravity, act upon its
particles.

35. It follows, from the preceding definition, that if the
centre of gravity of a body be fixed, the body will rest in any
position. For in every position the resultant of all the forces
arising from gravity passes through the fixed point, and being
met by the resistance of that point, equilibrium is preserved.
Hence, the centre of gravity of a body may be also defined as
that point about which the body will balance in every position.

36. Since the resultant may be always substituted in place of
its component forces, it follows that in considering the influence
of any weighty body in producing equilibrium, we may substitute a force equal to the weight of the body, and acting at its
centre of gravity.

37. *If a body suspended from any point be at rest, the centre
of gravity must lie in the vertical line drawn through this point.*
For there are two forces in equilibrium; viz., the weight of the
body which acts vertically, and at the centre of gravity, and
the reaction of the fixed point. By Principle I. these forces

must lie in the same straight line, and therefore the vertical line through the centre of gravity must pass through the fixed point.

Hence, if one point in any body be fixed, there are two positions, and two positions only, in which the body can rest, viz., when the centre of gravity is vertically over, or when it is vertically under the fixed point.

## 38. Centre of gravity determined experimentally.—

The principle established in the preceding article enables us to find the centre of gravity of any body by a very simple experiment. Take any point in the body, suspend it from this point, and when at rest draw the vertical or plumb-line through the point of suspension. By the preceding article, the centre of gravity must lie in this line. Take a second point in the body, and proceed as before. We have then another line in which also the centre of gravity must lie. The intersection of these lines will be the centre of gravity required; for it is the only point which lies in both lines.

When the body is a thin plate this method may be readily applied, as the vertical lines may be drawn upon the plate itself. If the body be a solid mass this method is very inconvenient, as in order to draw the vertical lines it becomes necessary to pierce the substance of the body. If the body be hollow like a bowl, or of an irregular shape like a chair, the inconvenience is somewhat less, as the vertical lines may sometimes be represented by means of wires or strings.

39. The position of the centre of gravity of any body depends, it will be seen, upon its shape; but not upon its shape only. Two bodies may have precisely the same shape, and also be of precisely the same size and weight, and yet their centres of gravity may not be in the same position. If both bodies are of uniform density, or if the density of each varies in the same way, in either case their centres of gravity will be in the same position. But if one be of uniform and the other of variable density, or if both being of variable density, the den-

sity of one varies in a different way from the other, then their centres of gravity will not be in the same position.

The position of the centre of gravity does not at all depend upon the absolute weight of the body. Two bodies may differ in weight, but if all their parts be similarly arranged, their centres of gravity will agree in position. Thus a bar of given length and of uniform density will balance about its middle point, whether made of wood or made of steel.

Hence, when density is supposed to be uniform, the only point to be considered in the determination of the centre of gravity is its shape, and it is on this supposition that the centre of gravity of certain geometrical figures is spoken of. By the centre of gravity of a triangle or a parallelogram is meant the centre of gravity of a thin plate of these forms respectively, and of uniform thickness and density.

### 40. Centre of gravity determined graphically.—The

geometrical centre of any perfectly symmetrical figure must also be the centre of gravity, for there can be no reason why the centre of gravity should fall upon one side of that point more than upon another. Hence, the centre of gravity of a line is its point of bisection : of a parallelogram, the intersection of its diagonals ; of a parallelopiped, the intersection of its diagonals ; of a circle and sphere, their centres.

A triangle is a figure which has no geometrical centre ; its centre of gravity, however, may be found in the following way. Let ABC be the given triangle. Bisect AB in D, and AC in E. Draw CD, BE. If we suppose the triangle to be made up of an in-finite number of lines parallel to AB, these lines will be bisected by CD. Consequently, the centre of gravity of each of these lines will be in the line CD, and therefore the centre of gravity of the triangle will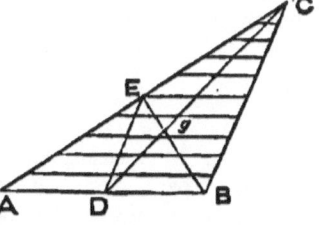
lie in the line CD. For similar reasons the centre of gravity must lie in the line BE. Therefore $g$, the intersection of these two lines, must be the centre of gravity required.*

* Also $gD = \frac{1}{3} CD$. For, join the points D, E. DE will be parallel to BC, and equal to half BC. Hence, the triangle $g$ED is equiangular with the triangle $g$BC, and the sides of the former are each half of the corresponding sides of the latter. Therefore $gD = \frac{1}{2} gC = \frac{1}{3} CD$. Hence, the

The centre of gravity of any four-sided figure may be determined by first drawing one of its diagonals, and joining the centres of gravity of the two triangles into which it is thus divided. As the centres of gravity of the two parts are in this line, the centre of gravity of the whole must be in this line. In like manner, by drawing the other diagonal, a second line may be found, in which also the centre of gravity must lie. The centre of gravity of the four-sided figure will be the intersection of these lines.

Thus, let ABCD be the given figure. Draw AC and BD. Let K, L, M, N be the centres of gravity of the triangles ACD, ACB, BDA, BDC respectively. Draw KL and MN intersecting in G. G is the centre of gravity required.

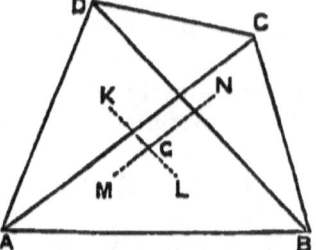

The centre of gravity of a five-sided figure may be found by dividing it into a triangle and a four-sided figure ; and by joining the centres of these two parts, one line is found in which the centre of gravity of the whole figure must lie. Dividing it similarly in another way, another such line may be found. The intersection of these two lines is the required centre of gravity.

By proceeding in a similar way the centre of gravity of any rectilinear figure may be determined graphically.

### 41. Centre of gravity determined by calculation.—

Whenever the weights and centres of gravity of the parts are known, the centre of gravity of the whole may be easily found by applying the principle of moments. (Art. 25.)

The following examples will illustrate the method :

*Ex.* 1. Two balls, weighing 8 and 10 oz. respectively, are connected by a weightless rod, the centres of the balls being 9 inches apart, find the distance of the centre of gravity from the centre of the heavier ball.

Let $x$ be the distance required. Then the entire weight is 18 oz. Taking the moments about the centre of the heavier ball, we have

$$18x = 8 \times 9 = 72,$$
$$\therefore \qquad x = 4 \text{ inches.}$$

centre of gravity of a triangle lies in the line joining any vertex with the bisection of the opposite side, at a distance from the side of one-third of this line.

*Ex.* 2. Two balls, weighing 12 and 7 oz. respectively, are connected by a rod of uniform thickness and density, weighing 5 oz., the distance between the centres of the balls is 16 inches, required the distance of the centre of gravity from the heavier weight.

Here we have three weights, whose magnitudes are respectively

$$12, \qquad 5, \qquad 7,$$

and the distances of their centres of gravity from the point of reference

$$0, \qquad 8, \qquad 16.$$

Hence, if $x$ be the distance required, we have

$$24x = 40 + 112 = 152,$$
$$\therefore \qquad x = 6\tfrac{1}{3}.$$

When the centres of gravity of the given weights do not lie in a straight line, the centre of gravity of the whole may be found by a mixed method, as in the following examples :

*Ex.* 1. To find the centre of gravity of three equal weights suspended from the angular points of a triangular plate, the plate being supposed to be without weight.

Let a weight P be suspended from each of the angular points A, B, C of the triangular plate ABC. Since these weights hang vertically, they may be regarded as parallel forces acting severally at the points of suspension. Bisect AB in D. Join DC. Let $DG = \tfrac{1}{3} DC = \tfrac{1}{2} GC$. The result- 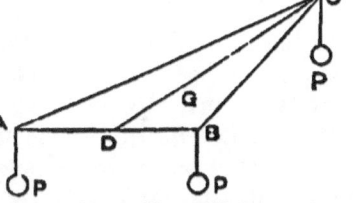 ant of the force P at A, and the force P at B, will be a force 2P acting at D. The resultant of 2P at D, and P at C, will pass through G ; for since $DG = \tfrac{1}{2} GC$,

$$2P \times DG = P \times GC ;$$

G is therefore the centre of gravity required.

By the preceding section, G is the centre of gravity of the triangle ABC, whence it appears that the centre of gravity of a triangle corresponds with that of three equal forces acting at its angular points.

*Ex.* 2. To find the centre of gravity of four weights, P, 3P, 7P, and 5P, suspended from the angular points of a square plate.

Let ABCD be the square plate, and let the weights P, 3P, 7P, and 5P be suspended severally from the points A, B, C, D.
Draw the diagonals, AC, BD.
Take N so that $CN = \frac{1}{8} AC$, and
M so that $DM = \frac{3}{8} BD$. Then
$AN = \frac{7}{8} AC$, and $BM = \frac{5}{8} BD$.

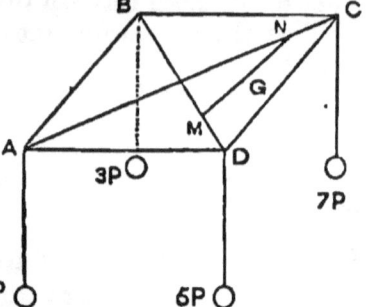

The resultant of P at A, and 7P at C, is a force 8P at N, since $7P \times CN = P \times AN$. And the resultant of 3P at B, and 5P at D is also a force 8P at M; since $5P \times DM = 3P \times BM$.
We have then in the place of the original forces two equal forces acting at M and N. Bisect MN at G, then the resultant of these equal forces will pass through G, and G is the centre of gravity required.

42. **Different kinds of equilibrium.**—A body at rest is said to be in *unstable* equilibrium, if after the slightest possible disturbance it moves away from its former position; in *stable* equilibrium, if it returns to its former position; and in *neutral* equilibrium if it does neither, but remains at rest in the new position into which it was brought by the disturbance.

When a body is suspended at any point, and the centre of gravity is vertically over the point of suspension, or as high as possible, the equilibrium is unstable. When the centre is vertically below the point of suspension, or as low as possible, the equilibrium is stable. If the point of suspension be the centre of gravity itself, the equilibrium is neutral.

A cone resting on its vertex would be a case of unstable equilibrium; when resting with its base on a horizontal plane, the equilibrium is stable; and when resting with its conical surface upon a horizontal plane, the equilibrium is neutral.

Stability of equilibrium is greater or less according as the angle is greater or less through which the body may be moved before it will cease to return to its original position. Thus, for instance, a book resting upright upon its edges is in stable equilibrium, because it may be slightly disturbed and yet will return to its position; but it is *more stable* when resting on its side, because it may be disturbed to a greater degree before it will cease to return to its original position of rest.

43. *A body placed with its base upon a plane surface will stand or fall, according as the vertical line through its centre of gravity falls within or without the base.*

Let ABCD be any body, whose base AB rests upon a plane surface. Let the centre of gravity be at $g$, so that the vertical line through $g$ falls between A and B. The weight of the body acting in this line is met by the resistance of the plane, and, consequently, will not cause the body to turn about either A or B.

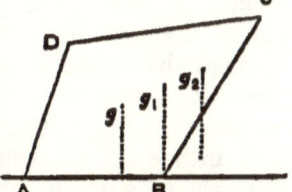

But if the centre of gravity be at $g_2$, so that the vertical line drawn through the centre of gravity falls without the base, the weight of the body, acting in this line, is not met by the re-action of the plane. There cannot, therefore, be equilibrium, but the body will turn over on to the side BC.

If the centre of gravity be at $g_1$, so that the vertical line drawn through it pass through one of the extremities of the base, the body will rest; for the weight acting in the vertical line is met by the resistance of the plane: a very slight disturbance towards the right hand will, however, cause the body to fall over.

44. **Measure of stability.**—If G be the centre of gravity of any body, and A an edge of the base, the body must be moved until GA becomes vertical; that is, through the angle GAN, before it will have any tendency to turn over. This

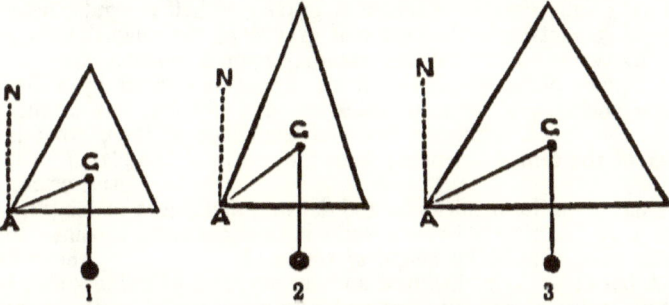

angle, therefore, may be taken as the measure of the stability of the body. Hence, other things being equal, the lower the centre of gravity the greater the stability, as is seen in figs. 1 and 2. Also, other things being equal, the greater the base the greater the stability, as is seen in figs. 2 and 3.

45. The statements of articles 43 and 44 apply equally, whether the plane upon which the body rests is horizontal or inclined, if only, in the case of the latter, the body be prevented by friction, or any other cause, from sliding down the plane. A body thus prevented from slipping will not, when placed upon an inclined plane, topple over, if the vertical line through the centre of gravity passes through the base, and will have no tendency to do so until the vertical line through the centre of gravity passes through the edge of the base.

If either of the figures represented in the preceding article were placed with the base AB upon an inclined plane, it would not topple over until the line GA were vertical; that is to say, it would not turn over if the inclination of the plane be less than the angle GAN.

## EXAMPLES.

1. Three heavy particles, whose weights are 4, 6, and 8 respectively, are placed at intervals of 9 inches each along a weightless rod, find the distance of the centre of gravity from the middle of the rod. Ans. 2 in.

2. If a triangular plate ABC form an isosceles triangle, whose base is AB, and whose height is 12 inches, find the distance of the centre of gravity from C, when weights 10, 10, and 40 lbs. are suspended from A, B, and C respectively; the plate being supposed to be without weight.
Ans. 4 in.

3. In the preceding example, if the plate weigh 20 lbs. what is the distance of the centre of gravity from C?        Ans. 5 in.

4. In example 2, if ABC be an equilateral triangle, whose side is 20 inches, find the distance of the centre of gravity from C.     Ans. 5·77.

5. Find geometrically the centre of gravity of half a regular hexagon.

6. Find geometrically the centre of gravity of half a regular octagon.

7. Find geometrically the centre of gravity of the quadrilateral figure formed by two isosceles triangles, standing upon a common base.

8. A square plate, whose side is 18 inches, is divided by a diagonal into two portions, of uniform thickness and density, but of different weights, the one weighing 12 lbs. and the other 38 lbs.; what is the distance of the centre of gravity from the centre of the plate?
Ans. 2·206 in.

9. A bar of uniform thickness and density, and 4 feet in length, has a weight of 10 lbs. attached to one end : it balances about a point 9 inches from that end, what is the weight of the bar?        Ans. 6 lbs.

10. A bar of uniform thickness and density, and weighing 5 lbs., has a weight of 10 lbs. attached to one end, and a weight of 12 lbs. from the other ; it balances about a point 4 inches from the middle, what is the length of the bar?        Ans. 9 feet.

# CHAPTER III.

## ON THE SIMPLE MACHINES.

46. A *machine* is an instrument, by the agency of which one force can resist or overcome another force not immediately opposite to it in direction.

The simplest of such instruments are sometimes denominated the *Mechanical Powers*, but are more fittingly termed the Simple Machines. These are—the lever, the wheel and axle, toothed wheels, the pulley, the inclined plane, the wedge, and the screw. These differ from each other more in their structure than in the principles of their operation; for when in equilibrium, the wheel and axle, the toothed wheel, and the pulley may be reduced to the lever; and the wedge and the screw are but modifications of the inclined plane.

The two forces which act upon either of these simple machines are, for the sake of distinction, called the *power* and the *weight*, the latter always denoting the force to be resisted or overcome.

47. By the *mechanical advantage* of any machine is meant the ratio of the weight to the power, when in equilibrium; thus, if a power of 2 lbs. sustain a weight of 30 lbs., the mechanical advantage is $30 \div 2$, or 15.

And hence, if W be the weight, P the power, and $a$ the advantage,
$$W = aP.$$

48. **The lever.**—The lever is an inflexible bar, capable of free motion about a fixed axis, called the fulcrum. Unless the contrary be stated, the lever is usually supposed to be without weight.

Levers are of three kinds, according to the relative position of the power, weight, and fulcrum.

When the fulcrum is between the power and the weight, the lever is of the first kind.

When the weight is between the fulcrum and the power, the lever is of the second kind.

When the power is between the fulcrum and the weight, the lever is of the third kind.

The beam of a balance is a lever of the first kind, an oar is a lever of the second kind, and the treadle of a lathe is a lever of the third kind.

Scissors are double levers of the first kind, nut-crackers are double levers of the second kind, and spring shears are double levers of the third kind.

### 49. Condition of equilibrium in the lever.

—According to what has been stated in Article 26, there will be equilibrium, if the resultant of the forces acting upon the lever passes through the fulcrum. Let P be the power and W the weight; let $a$ be the perpendicular distance of the fulcrum from the line in which P acts, and $b$ its distance from the line in which W acts. Then, whatever the directions of P and W, and whatever the form of the lever, their resultant will pass through the fulcrum, if

$$Pa = Wb.$$

Hence, if the power and the weight are the only forces acting upon the lever, that is to say, when the weight of the lever itself is disregarded, the condition of equilibrium in a lever

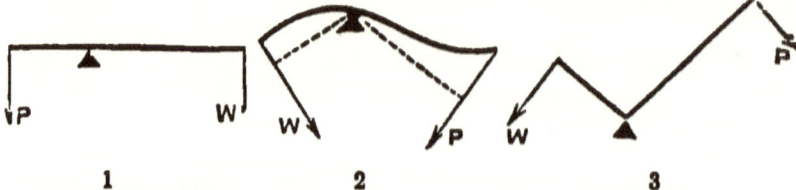

1　　　　　　　　2　　　　　　　　3

of any kind is, that *the power × its distance from the fulcrum = the weight × its distance from the fulcrum;* and, therefore, the

$$\text{mechanical advantage} = \frac{\text{dist. of power from fulcrum}}{\text{dist. of weight from fulcrum}}.$$

It is convenient to describe the perpendiculars drawn from the fulcrum to the directions of the power and the weight as the power's arm and the weight's arm respectively; and hence we may say, that when the weight of the lever is disregarded

$$P : W :: W\text{'s arm} : P\text{'s arm} ;$$

or, the mechanical advantage $= \dfrac{P\text{'s arm}}{W\text{'s arm}}.$

N.B. The arms of a lever are portions of the bar itself only when the forces act at right angles to the bar, as in figs. 1 and 3.

If the weight of the lever itself be regarded, the relation between the power and the weight may be found by means of the principle of moments; the sum of the moments of those forces which tend to turn the lever in one direction must be equal to the sum of those which tend to turn it in the contrary direction.

*Ex.* 1. A weight of 20 lbs. is suspended at a distance of 3 inches from the fulcrum of a straight lever of the first kind; where must a power of 4 lbs. be suspended, in order that there may be equilibrium, the weight of the lever being disregarded?

Let $x =$ the distance sought, then

$$4x = 20 \times 3 = 60,$$
$$x = 15 \text{ inches.}$$

*Ex.* 2. In a straight lever of the second kind, whose length is 18 inches, where must the weight be placed, if $P = 7$, when $W = 42$, the weight of the lever being disregarded?

Let $x$ be distance of W from fulcrum, then

$$42x = 7 \times 18 = 126,$$
$$x = 3 \text{ inches.}$$

*Ex.* 3. The arms of a straight lever of the first kind, of uniform thickness and density, are 7 in. and 3 in. respectively in length; a weight of 12 lbs. is suspended from the shorter arm, what power will support it, the weight of the lever being 4 lbs?

Since the lever is of uniform thickness and density, its centre of gravity is at half its length, and therefore at a distance of 2 in. from the fulcrum; and the weight of the lever tends to produce rotation in the same direction as the power.

Let P be the force required, then

$$P \times 7 + 4 \times 2 = 12 \times 3,$$
$$\text{or} \quad 7P + 8 = 36,$$
$$\therefore \quad P = 4 \text{ lbs.}$$

## 50. Pressure upon the fulcrum of a lever.—The pressure upon the fulcrum is in every case the resultant of the

forces acting upon the lever. In two cases this may be readily
found. First, when P and W are parallel forces and the weight
of the lever is disregarded. Then (Art. 13) the resultant equals
their sum or their difference, according as they act concurrently
or non-concurrently. Hence the pressure upon the fulcrum,

in levers of the first kind, is $P + W$,

in levers of the second kind, $W - P$,

in levers of the third kind, $P - W$.

Secondly, when P and W are both vertical, and the weight of
the lever is to be regarded.

The weight of the lever itself, being regarded, must then be
added or subtracted, according as the greater of the two weights
acts vertically downwards or vertically upwards. Thus, if $w =$
the weight of the lever, the pressure upon the fulcrum,

in levers of the first kind, is $P + W \pm w$,

in levers of the second kind, $W - P \pm w$,

in levers of the third kind, $P - W \pm w$.

In other cases the pressure upon the fulcrum must be found
by the application of the parallelogram of forces.


51. **Balances.**—Balances, mechanically considered, are simply
straight or bent levers of the first kind, with pans suspended
from the extremities of one or both of the arms.

The COMMON BALANCE is a balance with equal arms, and with
pans suspended from both its arms. Since the arms are equal,
a body placed in one of the pans will be balanced by an equal
weight in the other. If the scales be perfectly true, the body
and the weight will still balance, if they are made to change
places. This, then, is the most convenient way for testing a
common balance. If the body weighed is balanced by different
weights, according as it is placed in the one pan or the other,
the scales are shown to be false. The true weight of the body
is a mean proportional between the two false weights.*

* Let $w$ be the true weight of the body, and let $x$ and $y$ be the
unknown arms of the balance. When $w$ is placed in one pan, let it be
balanced by a weight of 16 lbs.; and when placed in the other, by a
weight of 12 lbs. Then, since the weight $w$ and 16, acting at the arms $x$
and $y$, were in equilibrium,    $wx = 16y$,
and since $w$ and 12, acting at the arms $y$ and $x$, were in equilibrium,
$$wy = 12x.$$
Multiplying together $w^2xy = 192xy$,
$$\therefore \quad w^2 = 192, \text{ or } w = \sqrt{192};$$
hence the true weight is a mean proportional between the two false weights.

Of balances with unequal arms, those most commonly used are the Steel-yard, the Danish Balance, and the Bent Lever Balance.

The STEEL-YARD is a lever of the first kind, with very unequal arms, provided with a moveable weight P, which may be suspended from any point in the longer arm. The longer arm is graduated, and the weight of any body suspended 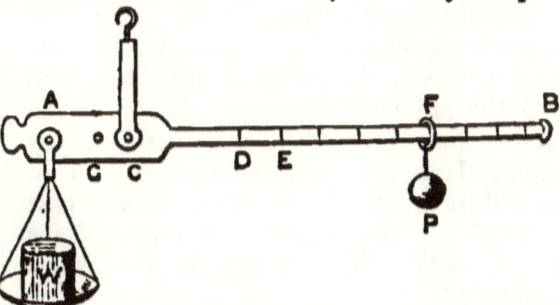 from the shorter arm is indicated by the point at which P must be placed in order to balance it.

The DANISH BALANCE.—This instrument consists of a straight bar AB, having a heavy knob A at one end, and at the other end B a hook, bearing a scale-pan. The fulcrum C is 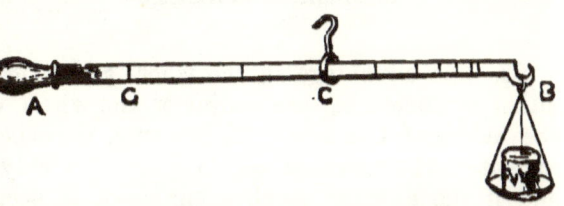 moveable, and the bar is so graduated, that the weight of any body placed in the scale-pan is determined by the position of C.

The BENT LEVER BALANCE.—This instrument, which is very convenient for determining expeditiously the weight of bodies within a moderate range, consists of a bent lever ACB, one end of which A bears an index, which moves over a graduated quadrant, and the other end B sustains the scale-pan. The thickness of the lever is greatly increased towards A, so that its centre of gravity shall be not far from A. The quadrant may be graduated experimentally, by placing different weights successively in the pan, and marking off the points at which A rests.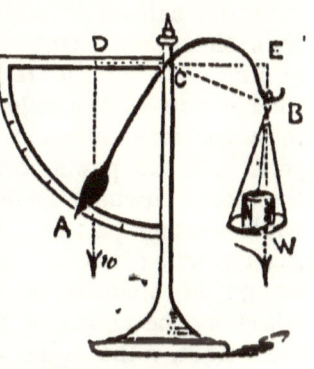

52. **The wheel and axle.**—The wheel and axle consists

of a cylinder or axle firmly fixed to a wheel, and having a common axis with it. The weight is attached to a cord passing round the axle, and the power to a cord passing round the wheel.

Let the figure represent a section of such a machine, where C is a point in the common axis, CA the radius of the wheel, and CB the radius of the axle. Let P be the power, and W the weight. These may be regarded as two parallel forces, and if in equilibrium, their resultant must pass through C, and therefore we must have $P \times AC = W \times BC$. Hence, in the wheel and axle

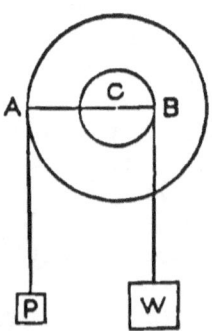

$P : W :: $ radius of axle : radius of wheel.
Whence,

$$\text{mechanical advantage} = \frac{\text{radius of wheel}}{\text{radius of axle}}.$$

53. The advantage in the wheel and axle may be increased either by increasing the radius of the wheel, or by diminishing the radius of the axle. If, however, the wheel be very greatly increased, the machine becomes too unwieldy to be serviceable, and if the axle be much diminished, it becomes too weak to sustain the weight. These difficulties in the way of an indefinite increase of the mechanical advantage are overcome by the simple device of a compound axle, one-half of which is of smaller radius than the other. One end of the cord sustaining the weight is wound round the thicker part of the axle, and the other end, in a contrary direction, round the thinner part. As the power descends, some part of the cord unwinds from the thinner axle, while another part is wound up around the thicker axle; but as more of the cord is wound up than is let out, the weight is raised by the action of the machine. Let P be the power, W the weight, $r$ the radius of wheel, $a$ the radius of thicker axle, and $b$ the radius of the thinner axle. Since the whole weight is supported by the two parts of the cord, the tension in the cord $= \frac{1}{2}W$.

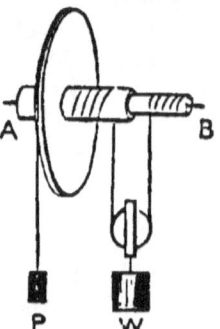

The power and the tension in the cord passing to the thinner axle both act on the one side of the axle, and the tension in the cord passing to the thicker axle acts on the other. There will be equilibrium, if the moment of the resultant of the former about a point in the axis equals the moment of the latter; and therefore, by Article 25, if the sum of the moments of the former equals the moment of the latter,

$$\therefore \qquad Pr + \tfrac{1}{2}Wb = \tfrac{1}{2}Wa,$$
$$\text{or} \qquad Pr = \tfrac{1}{2}W(a - b)$$
$$\therefore \qquad P : W :: \tfrac{1}{2}(a - b) : r.$$

Whence it appears that the mechanical advantage is equal to twice the radius of the wheel divided by the difference of the radii of the axles. Part of this advantage is owing to the introduction of the pulley, which, it will be presently seen, doubles the advantage of the machine. Consequently, the advantage of the wheel and axle alone is equal to the radius of the wheel divided by the difference of the radii of the axles; or the machine is equivalent to a simple wheel and axle, having an axle equal to the difference between the thicker and thinner parts of the compound axle.

54. **Toothed wheels.**—A toothed wheel is a circular plate of wood or metal, having its circumference indented or cut into equal teeth all the way round. If two such wheels, having their teeth of the same magnitude, and at the same distance apart, be so placed that a tooth of one may lie between any two of the other, then if one of them be turned round by any means, the other will be turned round also.

If the teeth be large, the amount and direction of the pressure which the one wheel exerts upon the other will, unless the teeth be made of a peculiar shape, vary considerably as the wheels revolve; but if the teeth be small in comparison with the size of the wheel, this variation may be disregarded, and the mutual pressure may, without any great error, be treated as constant in magnitude and direction.

55. *To find the ratio of the power and weight in toothed wheels, when the teeth are small.*

Let the power and the weight both act at axles of equal radius, and let $c$ be this radius. Let $a$ be the radius of P's wheel, and $b$ the radius of W's wheel.

Since the teeth are small, the pressure of the one wheel upon the other may be regarded as constant, and as acting in the direction of the common tangent to the two wheels. Let Q denote this pressure. Then, since P and Q are in equilibrium,

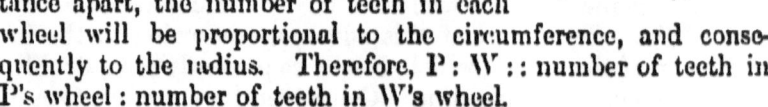

$$Pc = Qa.$$

Similarly,     $$Wc = Qb,$$

∴ multiplying crosswise, $PQbc = WQac,$

∴          $$Pb = Wa,$$

or,   $$P : W :: a : b.$$

But since the teeth in each wheel are of the same magnitude, and at the same distance apart, the number of teeth in each wheel will be proportional to the circumference, and consequently to the radius. Therefore, $P : W ::$ number of teeth in P's wheel : number of teeth in W's wheel.

Whence,

$$\text{mechanical advantage} = \frac{\text{number of teeth in W's wheel}}{\text{number of teeth in P's wheel}}.$$

**56. The pulley.**—A pulley consists of a small wheel, which moves freely about an axis, and allows a cord to pass over any part of its circumference. Unless it be otherwise stated, the wheel is supposed to revolve without friction, and the cord to be perfectly flexible.

No mechanical advantage is gained by a fixed pulley; for, as the tension in every part of the cord is the same, if a weight W be suspended at one extremity, an equal weight must be applied to the other to maintain equilibrium. Hence in this case,          $$P = W.$$

The effect of a fixed pulley is simply to change the direction of the force.

**57.** *To find the ratio of the power and weight in a single moveable pulley, when the cords are parallel.*

Let the weight W be attached to the moveable pulley B, and let B be sustained by a cord ABC, one extremity of which is fastened at A, and the other, after passing over the fixed pulley C, sustains the power P.

Let there be equilibrium: then the weight W is sustained by the tension in BA and the tension in BC; but since the cords are parallel, these tensions may be regarded as two parallel forces, and therefore W must equal their sum. But the tension of the cord is the same throughout, and is equal to P;

$$\therefore \quad W = 2P, \text{ or } P = \tfrac{1}{2}W.$$

58. *To find the ratio of the power and weight in a single moveable pulley, when the cords are not parallel.*

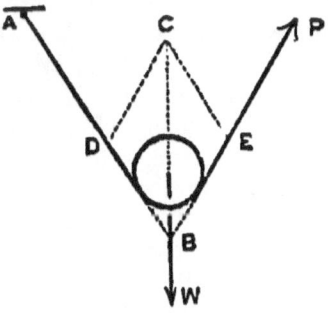

Let the directions of two parts of the cord meet in B. In the vertical line through B take any point C, and draw CD parallel to BE. Then, in the triangle CDB,

CB is parallel to W,
BD „ tension in BD,
DC „ tension in BE;

therefore the three forces are represented by the three sides of this triangle. But the tension in BD and the tension in BE are equal, being each equal to P, therefore BD = DC,

and 

$$P : W :: BD : BC ;$$

or,

$$\text{mechanical advantage} = \frac{DC}{BD}.$$

But one side of any triangle is less than the sum of the other two. Therefore, BC is less than BD + DC; that is, BC is less than 2BD. The mechanical advantage, therefore, is less than 2, and hence the advantage when the cords are inclined is less than when the cords are parallel.

59. The following combinations of pulleys are termed respectively, the first, second, and third system of pulleys.

In the first system of pulleys, each pulley hangs by a separate cord, one end of which is fastened to a fixed beam, and the other to the pulley above it. (See *fig. Art.* 60.)

In the second system of pulleys, the same cord passes round all the pulleys, which are arranged in two blocks, one of which is fixed, and the other bears the weight. (See *fig. Art.* 62.)

In the third system of pulleys, each cord is attached to the weight. (See *fig. Art.* 63.)

60. *To find the ratio of the power and weight in the first system of pulleys, when the weight of the pulleys is disregarded.*

Let ABC be three moveable pulleys, and let P and W be in equilibrium.

By Article 57,

tension in AB = $\frac{1}{2}$W,

tension in BC = $\frac{1}{2}$ tension in AB = $\frac{1}{4}$W,

tension in CF = $\frac{1}{2}$ tension in BC = $\frac{1}{8}$W.

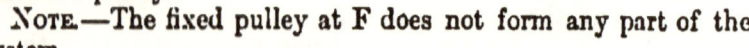

But, by Article 56, tension in CF = P, therefore, when there are three moveable pulleys, P = $\frac{1}{8}$W or W = 8P. Similarly, if there be four pulleys, W = 16P; if five, W = 32P, and so on. Hence, in the first system of pulleys, when the weight of the pulleys is disregarded, the weight is found by doubling the power as many times as there are pulleys, and the power is found by halving the weight as many times as there are pulleys.

NOTE.—The fixed pulley at F does not form any part of the system.

61. *To find the ratio of the power and weight in the first system of pulleys, when the weight of the pulleys is regarded.*

The total weight acting at each pulley is the tension of the string attached to the block, together with the weight of the block.

The entire weight at A is the weight W, together with the weight of the pulley A; one-half of this sum will give the tension in AB.

The entire weight at B is the tension in AB, together with the weight of the pulley B; one-half of this sum will give the tension in BC.

The entire weight at C is the tension in BC, together with the weight of the pulley C; one-half of this sum will give the tension in CF, which is equal to the power P.

Hence, when W is given to find P :—to W add the weight of the lowest pulley, and divide by 2 ; add the weight of the next pulley, and again divide by 2 ; repeat this process as many times as there are moveable pulleys, the result will give P.

If P be given, W is found by the inverse processes. Double P and subtract the weight of the highest pulley. Double again, and subtract the weight of the next pulley, and so on as many times as there are moveable pulleys.

*Ex.* 1. Three pulleys are arranged according to the first system, the lowest weighs 6 lbs., the next 5, and the highest 4; what power will sustain a weight of 1000 lbs. ?

$$1000$$
$$6$$
$$1006 \div 2 = 503$$
$$5$$
$$508 \div 2 = 254$$
$$4$$
$$258 \div 2 = 129 = P.$$

*Ex.* 2. Four pulleys are arranged according to the first system, the lowest weighs 4 lbs., the next 5, the next 3, and the highest 6 lbs. ; what weight will be sustained by a power of 43 lbs. ?

$$43 \times 2 = 86$$
$$6$$
$$80 \times 2 = 160$$
$$3$$
$$157 \times 2 = 314$$
$$5$$
$$309 \times 2 = 618$$
$$4$$
$$614 = W.$$

62. *To find the ratio of the power and weight in the second system of pulleys.*

Since there is but one cord, and P is attached to one extremity of it, the tension in every part is equal to P. Hence, if $n$ be the number of the portions of this cord in contact with the lower block, the weight supported will be $nP$; therefore in this system,

$$W = nP, \text{ or } P = \frac{1}{n}W.$$

If the weight of the lower block be $w$, the total weight supported is $W + w$; therefore, if the weight of the block be regarded,

$$W + w = nP, \text{ or } P = \frac{1}{n}(W + w).$$

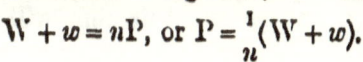

63. *To find the ratio of the power and weight in the third system of pulleys, when the weight of the pulleys is disregarded.*
Let there be three pulleys, A, B, C; then
tension in $Aa = P$,
tension in $Bb$ = pressure on A = 2P,
tension in $Cc$ = pressure on B = 4P.
But as the weight is supported by the tension in the three cords $Aa$, $Bb$, and $Cc$, and since the cords are parallel, W must equal the sum of the tensions; therefore,

$$W = P + 2P + 4P = 7P.$$

Similarly, if there be four pulleys,

$$W = P + 2P + 4P + 8P = 15P.$$

Hence, if as many terms of the series 1, 2, 4, 8, &c. as there are pulleys be added together, the sum multiplied by the power will give the weight; and, consequently, the weight divided by this sum will give the power.

*Ex.* 1. Five pulleys are arranged according to the third system, what weight will be supported by a power of 12 lbs., the weight of the pulleys being disregarded?

$$1 + 2 + 4 + 8 + 16 = 31,$$
$$W = 31 \times 12 = 372 \text{ lbs.}$$

*Ex.* 2. Six pulleys are arranged according to the third system, what power will support a weight of 504 lbs., the weight of the pulleys being disregarded?

$$1 + 2 + 4 + 8 + 16 + 32 = 63,$$
$$\therefore \quad P = 504 \div 63 = 8 \text{ lbs.}$$

64. *To find the ratio of the power and weight in the third system of pulleys, when the weight of the pulleys is regarded.*
The tension in $Aa$ (*fig. Art.* 63) equals the power. The tension in $Bb$ equals twice the power increased by the weight of the lowest pulley. The tension in $Cc$ equals twice the tension in $Bb$, increased by the weight of the second pulley. The weight W will be equal to the sum of the three tensions.
Hence generally, add together as many terms of the following series as there are pulleys; viz., the power, twice the power + weight of the lowest pulley, twice the preceding + the weight of the next pulley, and so on: this sum will be equal to the weight supported.

*Ex.* 1. Four pulleys, weighing respectively 6, 5, 4, and 3 lbs., are arranged according to the third system, the last mentioned being the lowest, what weight will be sustained by a power of 24 lbs. ?

$$P = 24$$
$$2 \times 24 + 3 = 51$$
$$2 \times 51 + 4 = 106$$
$$2 \times 106 + 5 = 217$$
$$398 \text{ lbs.} = W.$$

*Ex.* 2. With the same set of pulleys, find the power which will support a weight of 728 lbs.

Tension in first string = $P$
    „     second „ = $2P + 3$
    „     third   „ = $4P + 10$
    „     fourth „ = $8P + 25$
$$\therefore \quad 15P + 38 = W = 728$$
$$15P = 690$$
$$P = 46 \text{ lbs.}$$

65. **The inclined plane.**—*To find the ratio of the power and weight in the inclined plane, when the power acts parallel to the plane.*

Let W be a weight resting upon an inclined plane ABC, and supported by a power P acting parallel to the plane. Make $Ab = AB$, and $Ac = AC$, then the triangle $Abc$ is in all respects equal to the triangle ABC. In the triangle $Abc$,

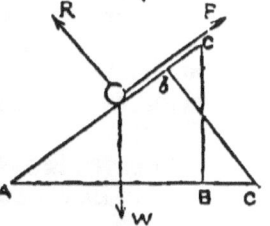

$bc$ is perpendicular to the direction of the power P,
Ac   „     „    the direction of gravity,
$Ab$   „     „    the re-action of the plane.

$\therefore$ by Principle IV. P : W :: $bc$ : Ac,
or               P : W :: BC : AC.
                         :: height of plane : length,
whence

$$\text{mechanical advantage} = \frac{\text{length of plane}}{\text{height}}.$$

In a similar manner the pressure of W on the plane may be determined. This is equal and opposite to the re-action of the plane. Let R denote this re-action,

$$R : W :: Ab : Ac,$$
$$:: AB : AC.$$

That is, the pressure on the plane is to W as the base of the plane is to its length.

66. *To find the ratio of the power and weight in the inclined plane, when the power acts horizontally.*

In this case W is kept at rest by three forces, viz., the force of gravity acting vertically, the power acting horizontally, and the re-action of the plane acting perpendicularly to the plane. The sides of the triangle ABC are severally perpendicular to the directions of these forces; viz.,

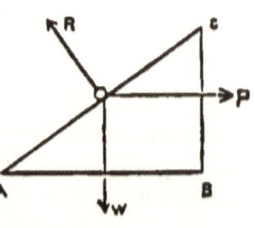

BC is perpendicular to the direction of P,
AB .............................................. W,
AC .............................................. R.

Therefore

$$P : W :: BC : AB :: \text{height of plane} : \text{base},$$

whence

$$\text{mechanical advantage} = \frac{\text{base of plane}}{\text{height}}.$$

Also, R : W :: AC : AB, or the pressure on the plane is to the weight as the length of the plane is to the base.

67. **The wedge.**—*To find the ratio of the power and the resistance in an isosceles wedge.*

Let ABC be the section of an isosceles wedge introduced into the cleft DFE, and let the points DE be similarly situated on the two sides of the wedge. The resistance on each side of the wedge will be the same, and if R be the total resistance, the resistances at D and E will each be ½R; and they act perpendicularly to the sides of the wedge. Let a power P act at the point H, the centre of the back of the wedge. The directions of these three forces when produced will meet in a point G; they may therefore be considered as three forces acting upon a point, and in equilibrium. The

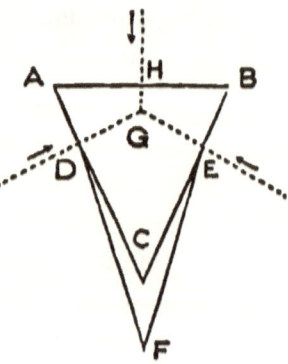

duced will meet in a point G; they may therefore be considered as three forces acting upon a point, and in equilibrium. The sides of the triangle ABC are severally perpendicular to the directions of the three forces, and therefore,

$$P : \tfrac{1}{2}R :: AB : AC,$$
$$\therefore \quad P : R :: \tfrac{1}{2}AB : AC.$$

That is, the power is to the total resistance as half the back of the wedge is to the side of the wedge.

68. **The screw.**—Let AB be a cylinder, and AD a rectangle, whose base AC is equal to the circumference of the cylinder. Let the side CD
be divided into any number of equal
parts. Join AG, and through the points
F, E, D, draw FH, EI, DK, parallel respectively to AG. Then if the rectangle
AD be wrapped round the cylinder, the
parallel lines AG, HF, &c. will trace out
the continued spiral line called the screw,
the distance between the threads of the
screw being equal to CG.

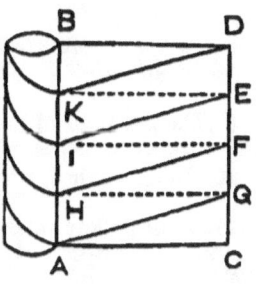

Hence any resistance to be overcome by a screw may be regarded as a weight resting on an inclined plane, whose base is equal to the circumference of the cylinder, and whose height is equal to the distance between the threads of the screw.

When the screw is used for mechanical purposes, the power is always applied in a direction perpendicular to the axis of the cylinder, and consequently parallel to the base of the inclined plane forming the screw line; and therefore

$$P : W :: \text{height of inclined plane} : \text{base};$$

that is, $P : W ::$ distance between the threads of the screw : circumference of the cylinder.

## EXAMPLES.

1. A bar 46 in. long, supposed to be without weight, is used as a lever of the first kind, where must the fulcrum be placed when a power 40 sustains a weight 190?*       Ans. 8 in. from the weight.
2. If the same bar be used as a lever of the second kind, where must the weight be placed?       Ans. 9·7 in. nearly from the fulcrum.

* Unless it be otherwise stated, the power and weight are supposed to act at right angles to the bar.

E

3. If the shorter arm of a lever of the first kind be 6 in., what is the length of the lever if P be 72 when W is 648 ?            Ans. 5 feet.

4. A bar 3 feet long, and weighing 6 lbs., is used as a lever of the first kind ; the shorter arm is 8 in.; what is P when W is 60 lbs. ?

Ans. 15 lbs.

5. With the same bar, what must be the length of the shorter arm, if P be 13 lbs. when W is 77 lbs. ?            Ans. 6 in.

6. The same bar is used as a lever of the second kind, the weight of the bar acting concurrently with W, what is P if W be 120 lbs. and W's arm be 6 in. ?            Ans. 23 lbs.

7. If the weight of the bar act concurrently with P, what will P be ?

Ans. 17 lbs.

8. In the simple wheel and axle the radius of the axle is 2 in., what must be the radius of the wheel if P be 10 when W is 160 ?

Ans. 2 ft. 8 in.

9. The two radii of a compound axle are 4 and 3 inches, the radius of the wheel is 30 inches, what is P when W is 720 lbs. ?            Ans. 12 lbs.

10. In Example 6, what is the pressure on the fulcrum ?

Ans. 103 lbs.

11. Four pulleys, whose weights, beginning with the highest, are 3, 4, 2, and 6 lbs. respectively, are arranged according to the first system, what power will sustain a weight of 462 lbs. ?            Ans. 32 lbs.

12. In the same system of pulleys, what weight will be sustained by a power of 62 lbs. ?            Ans. 942 lbs.

13. If four pulleys, weighing respectively 3, 7, 4, and 9 lbs., are arranged according to the first system, which is the most and which the least advantageous arrangement ?

14. What weight will be sustained by a power of 100 lbs. in the most advantageous arrangement, according to the first system, of four pulleys weighing respectively 4, 8, 4, and 10 lbs. ?            Ans. 1526 lbs.

15. With the same set of pulleys, what weight will be sustained by a power of 100 lbs. in the least advantageous arrangement ?

Ans. 1476 lbs.

16. If the same pulleys be arranged according to the third system, what power will sustain a weight of 1598 lbs. in the most advantageous arrangement ?            Ans. 100 lbs.

17. How many pulleys at least must be employed in the first system, in order that 1500 lbs. may be raised by 30 lbs., if each pulley weighs 4 lbs. ?            Ans. 6.

18. On an inclined plane, which rises 5 in 13,* what power, acting parallel to the plane, will support a weight of 91 lbs. ?            Ans. 35 lbs.

19. If 1000 lbs. be supported on an inclined plane rising 7 in 25, by a power acting parallel to the plane, what is the pressure on the plane ?

Ans. 960 lbs.

20. What horizontal force will sustain 1000 lbs. on an inclined plane rising 2 in 25 ?            Ans. 80·3 lbs. nearly.

* An inclined plane is said to rise 5 in 13, if when its height is 5 its *length* is 13.

21. What weight will be sustained by a horizontal force of 100 lbs. on an inclined plane rising 3 in 20 ?                    Ans. 659·1 lbs.

22. What is the pressure on the plane when a weight of 504 lbs. is sustained by a horizontal force on an inclined plane rising 16 in 65 ?
Ans. 520 lbs.

23. The diameter of a screw is 2 inches, and the distance between the threads is ¼ inch, what is the mechanical advantage ?*
Ans. 25½ nearly.

24. A bar of uniform thickness and density, 4 feet in length, is used as a lever of the first kind ; when the fulcrum is 7 inches from one end, a weight of 78 lbs. is balanced by a power of 10 lbs., what is the weight of the bar?                    Ans. 8 lbs.

69. *To find the mechanical advantage of any combination of machines.*

In a combination of machines, the weight of the first machine is the power of the second, the weight of the second the power of the third, and so on. Let $a$, $b$, $c$ be the separate advantages of three machines in combination. Let P and Q be the power and weight in the first, Q and R those in the second, R and W those in the third ; then

$$Q = aP,$$
$$R = bQ,$$
$$W = cR.$$

Multiplying together      $QRW = abc\ PQR,$
$$W = abc\ P.$$

Therefore $abc$ represents the mechanical advantage of the combination ; that is, the advantage of the combination is equal to the product of the separate advantages of the component machines.

As an example, take the case of the WEIGHING MACHINE. This is a combination of levers, so arranged as to furnish a convenient means for determining the weight of carriages and other heavy bodies. In a rectangular framework ABCD are placed four equal levers of the second order, having their fulcra

* The circumference of a circle is nearly ²²⁄₇ of the diameter ; more accurately 3·14159 times the diameter. It is usual to denote this ratio by the Greek letter $\pi$, and it will be useful to the student to remember the following expressions :

$$\text{circumference of circle} = 2\pi r,$$
$$\text{area of circle} \quad = \pi r^2,$$
$$\text{surface of sphere} \quad = 4\pi r^2,$$
$$\text{content of sphere} \quad = \tfrac{4}{3}\pi r^3,$$

where $r$ denotes the radius. See Newth's *Mathematical Examples*, pp. 74–79.

at A, B, C, and D, and their other extremities connected together at O, and also fastened there by a pivot to a lever KL. This lever is also of the second order, having its fulcrum at L, and is connected by a rod KM with a small balance MN. At E, F, G, and H are four pins, severally equidistant from A, B, C, and D. Upon these pins a platform (not represented in the figure) rests, and upon the platform the body to be weighed is placed. When used for weighing carriages the framework is sunk below the road, so that the platform may be on a level with the road. In place of the balance MN, a small steel-yard is frequently used.

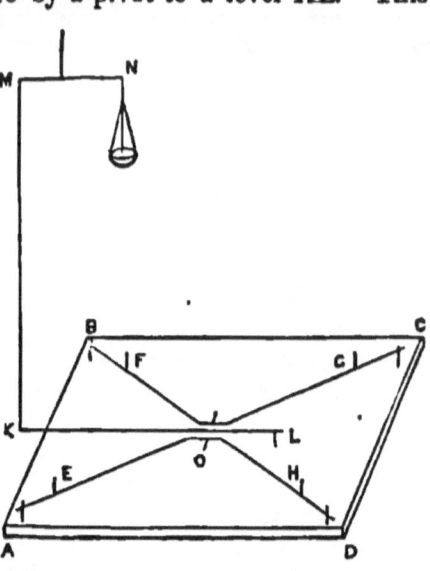

In determining the mechanical advantage of this machine it must be observed, that the four levers AO, BO, CO, DO are not in combination, but are simply a contrivance for the convenient support of the weight. The mechanical advantage is simply that of one of them. Hence, the instrument is equivalent only to a combination of three levers. Let A and $a$ be the lengths of the longer and shorter arms of the lever AO, B and $b$ those of the lever KL, and C and $c$ those of the lever MN; then,

$$\text{mechanical advantage} = \frac{A}{a} \times \frac{B}{b} \times \frac{C}{c},$$

$$= \frac{ABC}{abc}.$$

### EXAMPLES.

1. Four levers, whose arms are severally 10 and 2, 8 and 3, 15 and 1, 12 and 2, are used in combination, what is the mechanical advantage?     Ans. 1200.

2. In a common screw press, the diameter of the screw is 3 inches, the distance between the threads is ¼ inch, and the length of the lever is 3 feet, what is the mechanical advantage?     Ans. 905½ nearly.

3. Two pulleys are arranged as in the first system, the cords at each pulley are inclined at an angle of 60°, what is the mechanical advantage, the weight of the blocks being disregarded?     Ans. 3.

70. **The Principle of Virtual Velocities.**—If in any machine the power and weight be in equilibrium, and the machine be put in motion without disturbing the equilibrium, then the spaces described by the two forces are as the forces inversely, the spaces being measured in the direction of the forces respectively ; that is to say

$$\frac{\text{space described by P}}{\text{space described by W}} = \frac{W}{P}$$

This may be easily verified in the case of the simple machines.

Take for instance the wheel and axle. Let the wheel make an entire revolution, then the space through which P (*fig. Art.* 52) will move, will be that of the length of cord given out (or taken in) by the wheel; that is, will be equal to the circumference of the wheel : and the space through which W will move will be that of the length of cord taken in (or given out) by the axle ; that is, will be equal to the circumference of the axle.   Hence in this case the space described by P is to the space described by W as the circumference of the wheel is to the circumference of the axle.   But the circumferences of circles are as their radii ; hence

$$\frac{\text{space described by P}}{\text{space described by W}} = \frac{\text{radius of wheel}}{\text{radius of axle}}$$

But by *Art.* 52

$$\frac{W}{P} = \frac{\text{radius of wheel}}{\text{radius of axle}}$$

$$\therefore \quad \frac{\text{space described by P}}{\text{space described by W}} = \frac{W}{P}$$

If now, instead of causing the wheel to make an entire revolution, it make any part of a revolution, say one-tenth, the spaces described by P and W will be one-tenth of the circumference of the wheel, and one-tenth of the circumference of the axle, and these have the same ratio as before ; and so, whatever the distance through which the wheel may revolve, P's space is to W's space as the radius of the wheel is to the radius of the axle ; that is, as W is to P.

Again : take the first system of pulleys, and let there be three moveable pulleys as in *Art.* 60.   Let W be raised through any space, say 1 inch.   The lowest block will then be raised 1 inch, and 2 inches of cord will be given off from A ; the second block will then be raised 2 inches, and 4 inches of cord will be given off from B ; the third block will then be raised 4 inches,

and 8 inches of cord will be given off from C, and this consequently will be the space over which P will descend. Hence
the space described by P is 8 times that described by W; and
this is the ratio of W to P.

In like manner the principle may be verified with any number
of moveable pulleys.

Again: take the inclined plane when the power acts along
the plane. Let the weight move along the plane any distance,
and let AC (*fig. Art.* 65) represent this distance. Then the
distance through which P moves in its own direction is AC;
but W, acting vertically, moves in its own direction only through
the distance BC. Hence P's space is to W's space as AC is to
BC; that is, as the length of the plane is to the height, and
consequently as W is to P.

The student is recommended to verify the principle by similar
methods in the cases of the other simple machines.

71. By means of the principle of virtual velocities the
mechanical advantage of any machine may often be readily
found. If in any one instance the spaces described by P and
W in their own direction can be found, the ratio of these
spaces, being the same for all instances, gives us the ratio of
W to P, that is, gives us the mechanical advantage.

The following are examples:

*To find the mechanical advantage of the screw press.*

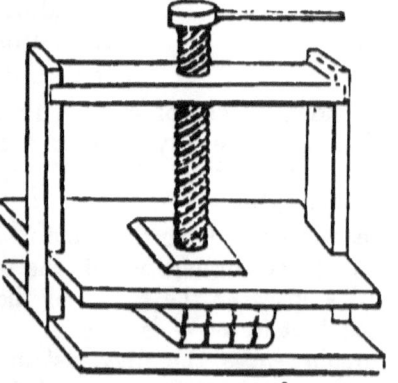

This instrument is a com-
bination of the screw and the
lever, the power being applied
at the extremity of a lever connected with the screw at right
angles with its axis. The fulcrum of the lever is at the axis
of the screw, consequently the
working length of the lever is
the distance from the axis of
the screw to the end of the
lever. If the lever make an
entire revolution, it is clear
that the screw will move forward through the distance between
the threads. Hence if $l$ be the length of the lever, and $d$ the
distance between the threads,

$$\text{mechanical advantage} = \frac{2\pi l}{d}.$$

*To find the mechanical advantage of the endless screw.*

This instrument is a combination of the screw and the wheel and axle. The screw is worked by a winch, and the threads of the screw press against the teeth formed on the circumference of the wheel. If the winch be turned round once one of the teeth of the wheel is moved forward; and hence if the winch be turned round as many times as there are teeth in the wheel, the wheel, and consequently the axle, will make one entire revolution. Hence if $n$ be the number of teeth, $l$ the length of the winch, and $r$ the radius of the axle, while P moves through $n$ times $2\pi l$, W moves through $2\pi r$, and therefore

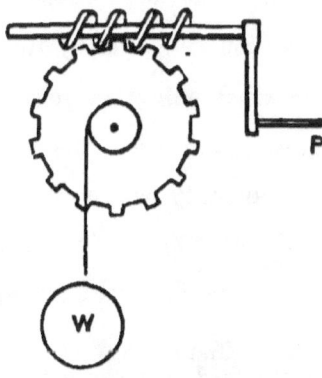

$$\text{mechanical advantage} = \frac{2\pi l n}{2\pi r} = \frac{l n}{r}$$

*To find the mechanical advantage of the Crane.*

This machine, used for raising and lowering heavy weights, is a combination of wheels and axles. A common form is represented in the figure. This consists of a winch and axle, and of two wheels and axles. The axle of the winch, and the axle of the smaller wheel, are furnished with teeth, as are also the two wheels. The power is applied at the handle of the winch, and the weight is sustained by a rope or chain passing round the axle of the second wheel.

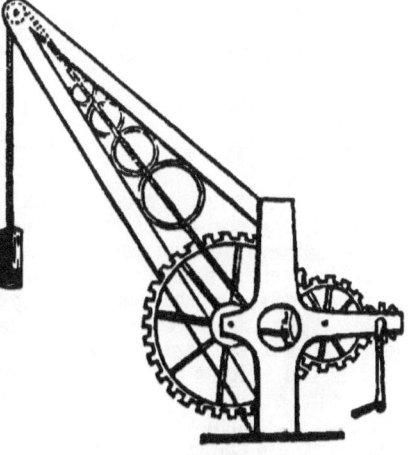

Let $r$ be the radius of the axle of the larger wheel, $N_1$ the number of teeth in the larger wheel, and $N_2$ the number in the smaller, $n_1$ the number of teeth in the axle of the smaller wheel, and $n_2$ the number in the axle of the winch, and let $l$ be the length of the winch.

Then, if the large wheel make one revolution, the weight will be moved over a distance $2\pi r$.

But if the large wheel move round once, the small wheel will move round $\dfrac{N_1}{n_1}$ times; and if the small wheel move round once, the winch will move round $\dfrac{N_2}{n_2}$ times. Hence if the large wheel move round once, the winch will move round $\dfrac{N_1}{n_1}\dfrac{N_2}{n_2}$ times. Consequently while W moves over $2\pi r$, P will move over $\dfrac{N_1}{n_1}\dfrac{N_2}{n_2}$ times $2\pi l$. Consequently

$$\text{mechanical advantage} = \frac{2\pi l\, N_1\, N_2}{2\pi r\, n_1\, n_2} = \frac{l\, N_1\, N_2}{r\, n_1\, n_2}$$

N.B.—The principle of virtual velocities, as stated above, is but the simplest and most elementary case of a comprehensive proposition applicable to any number of forces in equilibrium. In the cases given, two forces only are considered, the weight of the several parts of the machines being disregarded.

## EXAMPLES.

1. In an endless screw the length of the winch is 18 inches, there are 20 teeth in the wheel, and the radius of the axle is 4 inches, what weight will be sustained by a power of 50 lbs. ?  Ans. 4500 lbs.

2. In a screw press the length of the lever is 20 inches, what must be the distance between the threads that the advantage may be 500 ?  Ans. ·2513 inches.

3. In a crane the teeth in the two wheels are 36 and 18, and there are 6 teeth in each of the axles, the length of the winch is 24 inches, and the radius of the axle of the large wheel is 3 inches, what is the advantage ?  Ans. 144.

4.* Shew that in the wheel and axle, when the power and weight hang vertically, the position of their common centre of gravity is the same whatever the position of the weights.

5. If a weight be sustained on an inclined plane by another weight hanging freely, the two weights being connected by a cord passing over a pulley at the summit, shew that the common centre of gravity of the two weights lies in the same horizontal line in all positions of the weights.

6. If two weights connected by a cord passing over a pulley placed at the common summit rest upon a double inclined plane, shew from the principal of virtual velocities that the weights are to each other as the lengths of the planes on which they rest.

* In the solution of examples 4, 5, 6, the student must assume the similarity of equiangular triangles, or Euclid, b. vi. prop. iv.

# CHAPTER IV.

### ON THE LAWS OF MOTION AND THE MOTION OF FALLING BODIES.

72. When forces not in equilibrium act upon a body, motion must ensue. The swiftness with which the body moves is termed its velocity, and this may be either uniform or variable.

73. **Uniform velocity, and how measured.**—Velocity is uniform when the body describes *all* equal spaces in equal times. It is measured by the space described in a unit of time. It is customary to take a second for the unit of time, and a foot for the unit of measure; that is to say, velocity is measured by the number of feet described in one second. Thus, if a body moving with a uniform velocity have passed through 1000 feet in 8 seconds, its velocity is 1000 ÷ 8, or 125 ; and generally, if *s* be the space in feet through which a body has passed in *t* seconds, then if *v* denote the velocity,

$$v = \frac{s}{t} \quad \text{or} \quad s = vt.$$

Respecting this equation, it should be particularly noticed that its significance and its truth are based solely upon the understanding that velocity is measured in the way here described. All that it affirms is, that when a body moves uniformly, the number of feet described in any given number of seconds is equal to the number of feet described in one second multiplied by the number of seconds; and it is only as an abbreviation of this that it is allowable to say that

space = velocity × time.

74. **Variable velocity, and how measured.**—Velocity is variable when *all* equal spaces are *not* described in equal times. When a body in motion has a variable velocity, its velocity at any moment may be measured by the space which would be

described in a unit of time, if at that moment the velocity were to cease to vary.

**75. First law of motion.**—*A body in motion, not acted on by any external force, will continue to move in a straight line, and with a uniform velocity.*

This is equivalent to the assertion, that matter possesses no inherent power of changing the direction or the state of its motion.

The truth of this law must be decided by an appeal to experiment. The powers with which matter has been endowed can evidently be learnt only by observation. Every attempt to prove their existence by *a priori* demonstration will be found to assume in some way or other the property under consideration.

The first law of motion cannot, however, be established by any *direct* experiment, for the prescribed conditions can under no circumstances be fulfilled. Nowhere can we find a body which is not acted on by some external force. Every particle of matter is subject to a variety of external influences. Thus, if a ball be rolled along the ground, it is acted on by the attraction of all surrounding matter, by the resistance of the atmosphere, and by the force of friction; and it moves neither in a straight line nor with a uniform velocity. It is found, however, that the more we lessen the influence of external force, the more nearly does the motion become direct and uniform.

If a ball be rolled along a smooth pavement, it will move for a longer time, and in a line more nearly straight, than when thrown with the same velocity along a rough road; and still more so if rolled along a sheet of ice.

If a weight suspended by a thread from any point be made to oscillate, it will after a time come to rest. One of the external forces acting upon the body is the resistance of the air. If this be diminished by causing the body to move within the exhausted receiver of an air pump, the oscillation will continue for a longer period;—and the more perfect the vacuum the longer will the motion continue.

On a railway, after a train has acquired the desired velocity, it is no longer necessary for the engine to work, except so far as to overcome the effects of external forces, such as the friction of the rails and the resistance of the atmosphere.

The most convincing evidence of the truth of this law is found in the accordance of the consequences deducible from it with observed phenomena. It is impossible to doubt the correctness of a principle, upon the assumption of which the motions of the moon can be predicted with almost unerring certainty, and the time of an eclipse foretold within the fraction of a second.

**76. Uniform acceleration.**—It follows, from the first law of motion, that if a force continue to act upon a body already in motion, its velocity will continually change. This change may be either uniform or variable; uniform if the force be constant, variable if the force vary during the time of its action.

The change in the velocity resulting from the action of any force is uniform if in all equal periods of time there be an equal increase or an equal diminution in the velocity. The body is then said to be *uniformly accelerated*, and the acceleration is measured by the velocity added (or subtracted) in a unit of time. Thus, if a body move with a uniform acceleration of 32, it gains or loses in each second of time a velocity of 32 feet per second; that is, in one second it gains or loses a velocity of 32, in two seconds a velocity of 64, in three seconds a velocity of 96, and so on. And hence generally, if $f$ represent the acceleration, and $v$ be the velocity gained or lost in $t$ seconds,

$$v = ft.$$

And here we repeat the caution given in Article 73. The significance and truth of this equation are based solely upon the understanding that acceleration is measured in the way now described. All that it affirms is, that when a body moves under the action of any constant (that is, uniformly accelerating) force, the velocity generated in any given number of seconds equals the velocity generated in one second multiplied by the number of seconds; and it is only as an abbreviation of this that it is allowable to say that

velocity = acceleration × time.

**77. Second law of motion.**—*When a force acts upon a body in motion, the change of motion produced is the same, both in magnitude and direction, as if the force acted on the body at rest.*

Thus, if a body move along the line AB with such a velocity that it would describe the space AB in one second, and if, when it arrives at A, a force act upon it such as of itself to cause the body to pass from A to C in one second, then at the end of the second the body will be found at D; the change of motion represented by BD being the same in magnitude and direction as if the force had acted upon the body when at rest. 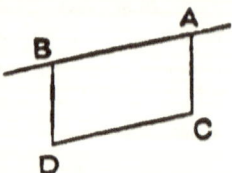 Each force produces its full effect in its own direction.

This law is proved by such experiments as the following :—

If a stone be dropped from the top of the mast, when a vessel is moving uniformly in any direction, and with any velocity, it will fall at the foot of the mast, just as it would if the vessel had been at rest.

If from any point a ball be let fall, and another ball be at the same instant projected forward horizontally with any velocity whatever, both balls will strike the ground at the same time. Here the ball at rest and the ball in motion are acted upon by the same vertical force, namely, the force of gravity, and both are caused to pass through the same vertical space in the same time.

If a person in a railway carriage throw a ball perpendicularly upwards, it will not fall towards the back of the carriage, but will drop into the hands of the individual who projected it.

78. **The third law of motion.**—*When force communicates motion to a body, the acceleration varies directly as the force and inversely as the mass.* *

This law is antecedently probable. We should expect, for instance, that if a certain mass be put in motion by the action of a certain force continued for a given time, say one second, and that if the same mass be put in motion by a force twice as great as the former, the velocity given to the body in the latter case will be twice as great as in the former; so also, that if the force be trebled, the velocity will be trebled; and generally that, the mass remaining the same, the velocity generated in one second will vary *directly* as the force.

Again, if different masses be put in motion by the same force,

* Mass, or quantity of matter in a body, may be measured by its weight; and it is as so measured that it is spoken of in the present and following sections.

continued for the same time, we should expect that the greater mass would have the smaller velocity, and that this diminution would be proportional to the magnitudes of the masses; that if, for instance, the one mass be double the other, the velocity generated will be only one-half as great; if treble, the velocity will be only one-third, and so on; or, in other words, when force communicates motion to a body, the acceleration varies *inversely* as the mass, if the force be constant.

These two statements are summed up in the single statement, that when force communicates motion to a body, the acceleration varies directly as the pressure and inversely as the mass.

The truth of this law is established by experiment.

Let different bodies of the same shape, and of the same material, but of different weight, be let fall from the same height, they will all reach the ground in the same time. The forces here causing motion are the weights of the bodies, but as are the weights, so also are the masses of the bodies, and hence any increase of the weight increases the mass in the same proportion. Since the different bodies fall through the same space in the same time, the acceleration when measured in the way already explained is, as will be presently shown, the same in all. These experiments, then, confirm the law, so far as it affirms that if the force and the mass change in the same proportion, the acceleration is the same.

In the preceding experiment, the condition that the bodies be of the same shape and material is rendered necessary by the pressure of the atmosphere, which might otherwise affect the several bodies unequally. If experiment be made in vacuo, this limitation is unnecessary, and it is then found that all bodies fall through the same space in the same time. A well-known experiment of this kind is that which is commonly described as the *guinea and feather experiment*. It is most conveniently performed by means of a long glass tube, closed at one end and furnished with a stopcock at the other. Let a guinea and a feather, or better still, a bullet and a pith ball, be placed in the tube, and let the air be exhausted. If the tube be turned quickly into a vertical position, both the enclosed articles will fall through the length of the tube in the same time.

In order to make experiments upon bodies of equal masses moved by unequal forces, or upon bodies of unequal masses moved by equal forces, a more elaborate apparatus is required. How this is done will be described in the following section.

### 79. Attwood's Machine. Experimental verification of the third law of motion.

—Attwood's Machine consists of two pillars, one of which is graduated, supporting a pulley arranged so as to work with as little friction as possible. An open ring A and a stage B slide along the graduated pillar, and by means of screws can be fixed at any part of it. P and Q are two equal cylindrical weights connected by a cord passing over the pulley. C is a pendulum beating seconds.

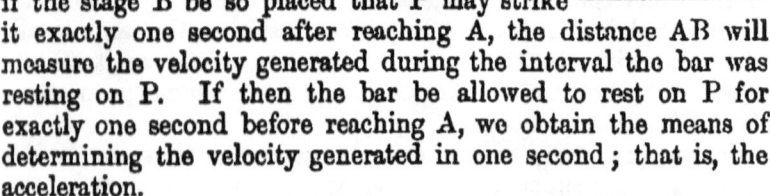

P and Q, being equal, will have of themselves no tendency to motion; but if a small bar be placed upon P, P will descend with an accelerated velocity until it reach the ring A, which, allowing P to pass through, but intercepting the bar, removes the cause of acceleration. P will then, in accordance with the first law of motion, move uniformly with the velocity it had acquired on reaching A; and if the stage B be so placed that P may strike it exactly one second after reaching A, the distance AB will measure the velocity generated during the interval the bar was resting on P. If then the bar be allowed to rest on P for exactly one second before reaching A, we obtain the means of determining the velocity generated in one second; that is, the acceleration.

The entire mass moved consists, it will be seen, of the bar and the two weights; the force causing motion is the weight of the bar only.

If now we arrange for a series of experiments in which the combined weights of P, Q, and the bar shall remain the same, but the weight of the bar vary, we shall find that the distance AB, which measures the velocity generated in one second, will vary in the same proportion as the weight of the bar. Thus if, for instance, in the first experiment P and Q are each $2\frac{3}{4}$ oz. and the bar $\frac{1}{2}$ oz., and in the second P and Q are each $2\frac{1}{2}$ oz. and the bar $1$ oz., we shall find that AB in the latter case is twice as great as in the former. In both experiments the mass moved is the same, namely, 6 oz., but the pressure in the latter case is double the pressure in the former.

If, again, we arrange for another series of experiments in which the bar is the same, but the weight of P and Q varies, we shall find that the distance AB varies inversely as the entire weight moved. Thus, let the bar weigh 1 oz., and in the first case let P and Q weigh $5\frac{1}{2}$ oz. each, and in the second $2\frac{1}{2}$ oz. each, the space AB will in the latter case be twice as great as the former. In both experiments the force causing motion is the same, namely, 1 oz., but the mass moved in the latter case is $2\frac{1}{2} + 2\frac{1}{2} + 1$, or 6 oz., or only one half of that moved in the former, which is $5\frac{1}{2} + 5\frac{1}{2} + 1$, or 12 oz.

80. **Momentum.**—The product of the mass of any body and the velocity with which it is moving is termed its *momentum.*

If $P_1$ and $P_2$ be forces acting upon bodies whose masses are $M_1$ and $M_2$, and if $f_1$, $f_2$ be the accelerations produced, then by the third law of motion

$$f_1 : f_2 :: \frac{P_1}{M_1} : \frac{P_2}{M_2}; \qquad \text{(i.)}$$

$$\text{or} \quad M_1 f_1 : M_2 f_2 :: P_1 : P_2. \quad \text{(ii.)}$$

Now if $v_1$ and $v_2$ be the velocities generated in the bodies by the forces in the time $t$, then $v_1 = f_1 t$, and $v_2 = f_2 t$. And since by the proportion just given

$$M_1 f_1 t : M_2 f_2 t :: P_1 : P_2,$$

it follows that

$$M_1 v_1 : M_2 v_2 :: P_1 : P_2; \quad \text{(iii.)}$$

or the momenta generated in equal times are as the forces producing them.

Hence the momenta generated in equal times by equal forces are equal, whether the masses be equal or unequal.

81. If the acceleration produced by gravity, or when a body falls freely, be known, then the third law of motion enables us to find the acceleration when any mass is set in motion by any force. Let $g$ denote the acceleration produced by gravity, and let $f$ denote the acceleration when a force P moves a mass M, both P and M being measured by weight. Then by hypothesis $g$ is the acceleration produced in the same mass M by a force M. And since the masses are equal, the accelerations are as the forces, and therefore

$$f : g :: P : M;$$

$$\text{or} \quad \frac{\text{acceler. produced by any force}}{\text{acceler. produced by gravity}} = \frac{\text{force measured by weight}}{\text{mass measured by weight}}.$$

## 82. Force of Gravity determined experimentally.—

One method of determining the value of $g$ is supplied by the equation just established. For by means of Attwood's machine we can find the value of $f$ experimentally for any values of P and M we please; and substituting these in the equation given above, we thence obtain the value of $g$.

Suppose, for instance, that when the bar is 1 oz., and the weights $7\frac{1}{2}$ oz. each, we find the distance AB to be 2 feet, then in this case the acceleration is 2 feet, the force 1 oz., and the mass moved is $1 + 7\frac{1}{2} + 7\frac{1}{2}$, or 16 oz. Hence

$$\frac{2}{g} = \frac{1}{16}; \qquad \text{or} \quad g = 32$$

By repeating the experiment with different weights, and by taking the average of all the results, a more reliable value of $g$ may be obtained than if we depended upon a single experiment.

The imperfections of Attwood's machine, arising from the friction of the wheels and other causes, do not enable us to do more than to obtain a rough approximation to the value of gravity in the way just described. Much greater accuracy is attained by means of experiments with pendulums. If $t$ denote the time of a vibration, $l$ the length of the pendulum, $g$ the force of gravity, and $\pi$ the ratio of the circumference of a circle to its diameter, then it can be shewn that

$$t = \pi \sqrt{\left(\frac{l}{g}\right)};$$

whence it follows that

$$g = \frac{l\pi^2}{t^2}.$$

The time of a single vibration of a pendulum can be determined with great accuracy by the simple device of observing the number of vibrations made in a considerable period, and then dividing the whole time by the number of vibrations. The length of the pendulum is found by measurement. By substituting these values when obtained in the equation just given, and by taking the average of a large number of results, it is found that at the sea level in the latitude of London the value of $g$ is 32·1908, or 32·2 nearly.

N.B.—By the length of the pendulum is here meant, not the length of the bar or bars of which the pendulum is composed, but the distance between the point of suspension and another point on the other side of the centre of gravity, so taken that, when the pendulum is suspended from it, the vibrations are made in the same time as before.

## EXAMPLES ($g-32$).

1. What acceleration will be given a body weighing 20 lbs. by a force of 5 lbs. ?  Ans. 8 ft. per sec.

2. If a body weighing 30 lbs. be moving with an acceleration of 12 feet per second, what is the force causing motion ?  Ans. $11\frac{1}{4}$ lbs.

3. A weight of 9 oz. is placed on a smooth horizontal table, and is joined by a cord passing over a pulley at the edge of the table to a weight of 1 oz. hanging freely, with what acceleration will the weights move ?  Ans. $3\cdot2$ ft.

4. Two weights of 9 and 7 oz. respectively joined by a cord passing over a fixed pulley hang freely, with what velocity will they be moving at the end of 3 seconds ?  Ans. 12 ft. per sec.

5. How long must a force of 1 lb. act upon a body whose weight is 16 lbs. that it may move with a velocity of 240 feet per second ?  Ans. 2 min.

6. What force acting uniformly for 10 seconds would give to a railway engine weighing 16 tons a velocity of 1 mile per minute ?  Ans. 4 t. 8 cwt.

7. A weight of 12 lbs. is placed upon a table, and the table descends with an acceleration of 8 feet per second, what is the pressure of the weight upon the table ?  Ans. 9 lbs.

The pressure of the weight upon the table is equal to the resistance of the table to the weight, and the force causing motion is the excess of the weight over the resistance of the table.

8. If in the preceding the table move upward with the same acceleration, what is the pressure of the weight upon the table ?  Ans. 15 lbs.

9. Two bodies whose weight are as 5 to 2 are moved by forces whose ratio is 2 to 5, compare the velocities generated in $t$ seconds.

Ans. As 4 : 25.

83. **Unit of force.**—The unit of force is the force which in a unit of time will generate in a unit of mass a unit of velocity.

In England the unit of time is one second, the unit of mass one pound, the unit of space one foot, and consequently the unit of velocity a velocity of one foot per second. Hence the unit of force is the force which in one second will generate a velocity of one foot per second in a mass whose weight is one pound. If P denote this unit, then its value may be found from the equation. (Art. 81.)

$$\frac{1}{g} = \frac{P}{1 \text{ lb.}} \text{ or } P = \frac{1 \text{ lb.}}{g};$$

or the unit of force is very nearly the force exerted by a weight of half an ounce.

In France the unit of mass is one gramme, and the unit of space is one centimetre ; and hence the unit of force is the force which in one second will generate a velocity of one centimetre

F

per second in a mass whose weight is one gramme. This unit is termed a *dyne*. The value of $g$ in centimetres is 981, and hence a dyne is $\frac{1}{981}$ of a gramme.

A convenient mode of designating these units of force is by the initial letters of the respective units of space, mass, and time. Thus the English unit may be designated the F.P.S. (foot—pound—second) unit; and the dyne, or French unit, the C.G.S. (centimetre—gramme—second) unit.

It will be seen from the above that using the F.P.S. unit $g$ or 32·2 is the number of such units exerted by a weight of 1 lb.; and hence if M denote the number of pounds in any mass, M$g$ is the number of units of force exerted by gravity upon that mass.

Similarly, if we use the C.G.S. unit, $g$ or 981 is the number of such units exerted by a weight of 1 gramme; and hence if M denote the number of grammes in any mass, M$g$ is as before the number of units of force exerted by gravity upon that mass.

**84. Motion of a body moving from rest under the action of a uniformly accelerating force.**—It will be found, by experiments with Attwood's machine, that the space described in one second is equal to one-half of the acceleration. Thus, if the force generates a velocity of 20 ft. per second, the space described in one second will be 10 ft.; and generally, if $f$ be the velocity generated per second, the space described in one second will be $\frac{1}{2}f$.

At the expiration of one second, the body will have acquired a velocity $f$, and, therefore, during the 2nd second the force is acting on a body in motion. By the second law of motion, each cause of motion will produce its full effect in its own direction. In the present case, both are acting in the same direction, and hence the entire space described in the 2nd second will be the sum of the separate spaces due to the velocity acquired and the action of the force. But the space due to the velocity acquired $=f$, and the space due to the action of the force $=\frac{1}{2}f$, therefore the space described in the 2nd second $= f + \frac{1}{2}f = \frac{1}{2}f \times 3$.

The velocity acquired in two seconds is $2f$. Hence, in the 3rd second, we have the space due to the acquired velocity $= 2f$, and space due to the action of the force $= \frac{1}{2}f$, therefore the space described in the 3rd second $= 2f + \frac{1}{2}f = \frac{1}{2}f \times 5$.

In like manner it will be found, that the space described in the 4th second $= \frac{1}{2} f \times 7$; in the 5th, $\frac{1}{2} f \times 9$; and so generally, *the space described in the nth second equals half the acceleration multiplied by the nth odd number.*

By adding together the spaces described in the 1st and 2nd second, we obtain the space described in two seconds; adding to this the space described in the 3rd second, we obtain the space described in 3 seconds, and so on. Hence,

space described in 2 seconds $= \frac{1}{2} f \times$ 4,

„    „    3 „    $= \frac{1}{2} f \times$ 9,

„    „    4 „    $= \frac{1}{2} f \times 16$;

and generally, *the space described in any time equals half the acceleration multiplied by the square of the number of seconds;* or, as it is commonly written,

$$s = \frac{1}{2} f t^2.$$

By Art. 76, $v = ft$, and therefore, $v^2 = f^2 t^2$. But, from the preceding, $2fs = f^2 t^2$. Hence,

$$v^2 = 2 fs,$$

or *the square of the velocity equals twice the acceleration multiplied by the space.*

The three equations, $v = ft$, $s = \frac{1}{2} f t^2$, and $v^2 = 2fs$, express the relation between each pair of the three quantities; viz., the space described, the velocity acquired, and the time when a body moves from rest under the action of a constant force. If any one of these quantities be known, either of the other two may, by means of one or other of these equations, be found directly. When gravity is the force considered, these equations are written $v = gt$, $s = \frac{1}{2} g t^2$, and $v^2 = 2gs$.

*Ex.* 1. A stone falls from rest under the action of gravity, find the space described in 5 seconds.

In the formula $s = \frac{1}{2} g t^2$, make $t = 5$, then

$$s = 16 \quad \times \quad 25 = 400 \text{ feet.}$$

*Ex.* 2. Find the velocity which a stone will acquire by falling through 1600 feet.

In the formula $v^2 = 2gs$, make $s = 1600$, then

$$v^2 = 64 \times 1600,$$
$$= 102400,$$
$$\therefore v = 320.$$

68 ON THE LAWS OF MOTION

*Ex.* 3. How long must a body fall, under the action of gravity, to acquire a velocity of 128 feet per second?

In the formula $v = gt$, make $v = 128$, then
$$128 = 32 \times t,$$
$$\therefore \quad t = 128 \div 32 = 4 \text{ seconds.}$$

*Ex.* 4. How far must a body fall, under the action of gravity, to acquire a velocity of 96?

In the formula $v^2 = 2gs$, make $v = 96$, then
$$(96)^2 = 64 \times s,$$
$$\therefore \quad s = 144.$$

*Ex.* 5. A weight of 9 ounces draws a weight of 7 ounces over a fixed pulley, find the space described from rest in $t$ seconds, neglecting the inertia of the pulley.

The pressure causing motion is $9 - 7 = 2$ ounces, and the weight moved is $9 + 7 = 16$ ounces, therefore (Art. 81),

$$\text{the acceleration} = \frac{2g}{16} = 4.$$

Substituting this in the general formula $s = \frac{1}{2} ft^2$, we have

$$s = 2t^2.$$

Hence, in one second the space described is 2 feet, in two seconds 8 feet, in three seconds 18 feet, and so on.

*Ex.* 6. A weight of 9 lbs. is drawn along a smooth horizontal table by a weight of 1 lb. hanging vertically by a string passing over a pulley at the edge of the table; find the space described from rest in 3 seconds, the velocity acquired in 4 seconds, and the space described in the 5th second.

The pressure causing motion is 1 lb., the weight moved is $9 + 1 = 10$ lbs. Therefore

$$\text{acceleration} = \frac{g}{10} = 3 \cdot 2,$$

space described in 3 seconds $= 1 \cdot 6 \times 9 = 14 \cdot 4$ ft.,
velocity acquired in 4 seconds $= 3 \cdot 2 \times 4 = 12 \cdot 8$.

The space described in the 5th second is $\frac{1}{2} f$ multiplied by the 5th odd number, or 9; $\therefore$ space described in 5th second is $1 \cdot 6 + 9 = 14 \cdot 4$ feet.

EXAMPLES ($g = 32$).

If a body fall from rest under the action of gravity, find

(1.) The space described in the 21st second.  Ans. 656 ft.
(2.) The velocity acquired in 8 seconds.  Ans. 256.
(3.) The space described in 6 seconds.  Ans. 576 ft.
(4.) The time of acquiring a velocity 160.  Ans. 5 sec.
(5.) The space described in acquiring a velocity 224.  Ans. 784 ft.
(6.) The time of describing 400 ft.  Ans. 5 sec.
(7.) The velocity acquired in describing 576 ft.  Ans. 192.

8. A weight of 8½ lbs. descends drawing up another of 7½ lbs. over fixed pulley, find the space described in 5 seconds.  Ans. 25 ft.

9. A weight of 13 lbs. is drawn along a smooth horizontal table by a weight of 3 lbs. hanging vertically, find the time of describing 147 ft.  Ans. 7 sec.

## 85. Motion of a body projected vertically upwards.—

If a body be projected vertically upwards, and be allowed to return to the point from which it was projected, then, disregarding the resistance of the atmosphere, the following two laws hold good :

First. The times of ascent and descent are equal.

Secondly. The velocities of ascent and descent are equal at equal altitudes.

From the latter, it follows directly, that on returning to the point of projection, the body is moving downwards with the same velocity as that with which it was projected upwards.

By the aid of these two laws various problems relating to the motion of a body projected vertically upwards are reduced to simpler questions of a body falling freely from rest. For instance, if the time which elapsed between the projection of a body and its return be known, half of this is the time of descent; and the question, How high did the body rise? resolves itself into the simpler question, How far will a body fall from rest in the time thus given?

Again, if the velocity of projection be given, and the question be, How long did the body rise? we have only to ask, How long must a body fall from rest to acquire such a velocity?

Or, thirdly, if the velocity of projection be given, and the question be, How high did the body rise? we can substitute for this the question, How far must a body fall from rest to acquire the given velocity?

The following are examples in illustration :

*Ex.* 1. A body is projected vertically upwards with a velocity of 100 feet per second, to find how high it will rise.

By Art 84, $v^2 = 2gs$, or $s = \dfrac{v^2}{2g}$ ; therefore, in this case,

$$s = \frac{10000}{64} = 156 \cdot 25.$$

*Ex.* 2. A body is projected vertically upwards with a velocity of 160 feet per second, to find how long it will continue to ascend.

By Art. 84, $v = gt$, or $t = \dfrac{v}{g}$; therefore, in this case,

$$t = \frac{160}{32} = 5 \text{ seconds.}$$

*Ex.* 3. A stone is projected vertically upwards, and returns to the same spot after an interval of 12 seconds, find the velocity of projection, and the height to which the body has risen.

The velocity of projection is equal to the velocity acquired by the body during the time of descent, or 6 seconds; and therefore, if V be the velocity of projection, since $v = gt$,

$$V = 32 \times 6 = 192 \text{ feet per second.}$$

The height to which the body has risen is equal to the distance through which it falls from rest during 6 seconds; that is,

$$\text{the height required} = \tfrac{1}{2}g \times 6^2$$
$$= 16 \times 36 = 576 \text{ feet.}$$

86. Another class of problems relating to the motion of a body projected vertically upwards is that which requires us to find either the velocity or the position of the body at a given time. These are readily solved by aid of the second law of motion. For since the direction of the force is exactly opposite to that of the velocity of projection, it follows that

*The velocity at any instant equals the velocity of projection diminished by the velocity generated by the force of gravity during the time of motion; and*

*The space described in any time equals the space due to the velocity of projection diminished by the space due to the action of gravity.*

The following are examples in illustration :

*Ex.* 1. A ball is projected vertically upwards with a velocity of 200 feet per second, what is its velocity at the end of 3 seconds? The velocity generated by gravity in 3 seconds is $32 \times 3$, or 96 ; therefore,

$$\text{velocity required} = 200 - 96 = 104.$$

*Ex.* 2. A ball is projected vertically upwards with a velocity of 120 feet per second, at what height will it be at the end of 4 seconds? The space due to the velocity of projection is (Art. 73) $120 \times 4$, or 480 feet. The space due to gravity in 4 seconds is $16 \times 4^2$, or 256 feet. Therefore,

$$\text{height required} = 480 - 256 = 224.$$

*Ex.* 3. A ball is projected vertically upwards with a velocity of 160 feet per second, and two seconds afterwards another ball is projected in the same direction with a velocity 224, when, and at what height, will the balls meet? Let the balls meet $x$ seconds after the projection of the second ball, then $x + 2$ is the time the first ball has been moving. The space described by the first ball in $x + 2$ seconds is

$$160 (x + 2) - 16 (x + 2)^2.$$

Similarly, the space described by the second ball in $x$ seconds is

$$224x - 16x^2.$$

But since the balls meet, these two quantities must be equal ; $\therefore$

$$224x - 16x^2 = 160 (x + 2) - 16 (x + 2)^2$$
$$\therefore \quad 128x = 256,$$
$$\therefore \quad x = 2 ;$$

or the balls meet 2 sec. after the projection of the second ball. To find the height, substitute 2 for $x$ in either of the expressions for the space described by the balls ; *e.g.* $224x - 16x^2$,

$$\text{height} = 224 \times 2 - 16 \times 4 = 384.$$

The laws stated in this section are formally deduced in the following sections.*

87. *To find the velocity acquired and the space described in*

_____
* These may be omitted on a first reading.

*a given time when a body is projected with a given velocity, and is acted on in the same direction by a constant force. Let V be the velocity of projection, f the acceleration due to the force, and t the time.*

From the second law of motion, it follows that the velocity will equal the velocity of projection + the velocity generated by the force; and, therefore, if $v$ denote the velocity required,

$$v = V + ft.$$

And similarly, the space described = the space due to the velocity of projection + the space due to the action of the force. The space described by a uniform velocity V in $t$ seconds = $Vt$, and the space due to the action of the force in $t$ seconds = $\frac{1}{2} ft^2$. Therefore

$$s = Vt + \tfrac{1}{2} ft^2.$$

Squaring both sides of the first equation, we have

$$v^2 = V^2 + 2Vft + f^2 t^2$$
$$= V^2 + 2f(Vt + \tfrac{1}{2} ft^2)$$
$$= V^2 + 2fs.$$

88. If the body is projected in the direction opposite to that in which the force acts, the expressions deduced as in the preceding article will be

$$v = V - ft,$$
$$s = Vt - \tfrac{1}{2} ft^2$$
$$v^2 = V^2 - 2fs.$$

89. *To find the height to which a body will rise when projected vertically with a given velocity* V.

Here the direction of V is opposite to that of the force; and hence, by the preceding article, substituting for $f$, the acceleration due to gravity, namely, $g$,

$$v^2 = V^2 - 2gs.$$

But when the body has attained the highest point, its velocity at that instant will = o, and therefore if $s =$ the height of this point,

$$o = V^2 - 2gs.$$
$$\therefore \qquad 2gs = V^2.$$

By comparing this result with the expression $v^2 = 2gs$ in Art. 84, we see that the height to which the body will rise is the same as the distance through which it must fall to acquire the velocity of projection.

90. *To find the time during which a body will rise when projected vertically with a given velocity* V.

Since at the instant of attaining the greatest height $v = 0$, then (Art. 88)

$$V - gt = 0$$
$$\therefore \quad gt = V$$
$$\therefore \quad t = \frac{V}{g}.$$

From the general formula $v = gt$, it follows that $\frac{V}{g}$ is also the time during which a body must fall to acquire the velocity V; and hence, from this and the preceding section, we see that on its return to the point whence it was projected, the body will have a velocity equal to the velocity of projection, and that the times of descent and ascent will be equal.

And further, since any point in the upward path of a projectile may be taken as the point of projection, if we suppose it to be projected thence with a velocity equal to that which it has then acquired, it follows that the velocity of the projectile at any altitude is the same both in its upward and downward course.

91. *If a body move under the action of a constant force, the difference of the squares of the velocities at any two points in its path is equal to the square of the velocity acquired by moving from rest under the action of the force through the intervening distance.*

Let $v_1$ and $v_2$ be the velocities at the two points, and $s_1$ and $s_2$ the distance of the points from the starting-point, then (Arts. 87, 88)

$$v_1^2 = V^2 \pm 2fs_1$$
$$v_2^2 = V^2 \pm 2fs_2$$
$$\therefore \quad v_1^2 - v_2^2 = \pm 2f(s_1 - s_2).$$

If $s$ be the distance between the two points $s = s_1 - s_2$ or $s_2 - s_1$, according as $s_1$ or $s_2$ is the greater, and if $u$ be the velocity acquired in moving from rest over the distance $s$ under the action of the force $u^2 = 2fs$, therefore

$$v_1^2 - v_2^2 = \pm 2fs = \pm u^2.$$

The following are examples of the application of this result:

*Ex.* 1. The velocity of a falling stone has increased from 20 to 30 feet per second, through what space has it moved in the interval?

Substituting for $f$ the acceleration due to gravity, or 32, we have

$$900 - 400 = 64s$$

$$\therefore \quad s = \frac{500}{64} = 7\tfrac{13}{16} \text{ ft.}$$

*Ex.* 2. A falling stone has a velocity of 20, what will be its velocity after it has fallen 6 feet further?

Let $v$ be the velocity, then, since $v$ will be greater than 20,

$$v^2 - 20^2 = 64 \times 6$$

$$\therefore \quad v^2 = 400 + 384 = 784$$

$$\therefore \quad v = 28.$$

*Ex.* 3. A stone moving vertically upward has a velocity of 40, what will be its velocity after rising 10 feet further?

Let $v$ be the velocity, then, since $v$ will be less than 40,

$$40^2 - v^2 = 64 \times 10$$

$$\therefore \quad v^2 = 1600 - 640 = 960$$

$$\therefore \quad v = 31 \text{ nearly.}$$

*Ex.* 4. The velocity of a body, moving under the action of a constant force over a distance of 12 feet, increases from 16 to 18 feet per second, what is the acceleration?

Let $f$ be the acceleration, then

$$18^2 - 16^2 = 2f \times 12$$

$$\therefore \quad 24f = 324 - 256$$

$$\therefore \quad f = 2\tfrac{5}{6} \text{ ft.}$$

## EXAMPLES.

1. A ball is projected vertically upwards with a velocity of 80 feet per second; how high will it rise in 2 seconds?    Ans. 96 feet.

2. A ball is projected vertically upwards with a velocity of 192 feet per second; after what interval will it return to the point of projection?    Ans. 12 sec.

3. A ball is projected vertically upwards with a velocity of 80 feet per second; what is its greatest elevation?    Ans. 100 feet.

4. A ball is projected vertically downwards with a velocity of 70 feet per second; what is the space described in 6 seconds?    Ans. 996 feet.

5. A ball is projected horizontally with a velocity of 60 feet per second; what is the distance of the ball from the point of projection at the end of 5 seconds?    Ans. 500 feet.

6. A ball is projected vertically upwards with a velocity of 1000 feet per second, and 3 seconds afterwards another ball is projected in the same direction with a velocity 1856; at what time after the projection of the second ball will the balls meet?    Ans. 3 sec.

7. If in the preceding question the balls had been projected downwards, when would they have met ?
Ans. 4⅝ sec. after the projection of the second ball.

8. A ball is projected horizontally with a velocity of 40 feet per second ; from the same point, and at the same instant, a second ball is let fall ; what distance will the balls be apart in 8 seconds ?  Ans. 320 feet.

9. Two balls are projected at the same time from the same point, one horizontally, with a velocity 60, and the other vertically downwards, with a velocity 20 ; what will be the distance of the balls from each other in 5 seconds ?  .  Ans. 316·23 feet nearly.

10. A falling stone increases its velocity 8 ft. per second by falling through 12 ft., what was its velocity at the first point of observation, and supposing it to have moved from rest, how far had it previously fallen ?
Ans. 44 ft. per second, 30¼ ft.

11. A stone rising vertically upward through a distance of 10 ft. loses in velocity 8 ft. per second, what was its velocity at the first point of observation, and how high will it rise from the second point ?
Ans. 44 ft. per second, 20¼ ft.

12. The velocity of a body when moving under the action of a constant force through a certain distance changes from 48 to 52 ft. per second, what velocity would it acquire if it moved from rest over the same distance ?  Ans. 20 ft. per second.

13. A body is projected along a rough horizontal plane with a velocity of 60 ft. per second, how soon will it be brought to rest, supposing the effect of friction to be a retarding force equal to 1–160th of the weight of the body ?  Ans. 5 min.

14. A body is projected along a rough horizontal plane with a velocity of 40 ft. per second, is brought to rest in 1½ min., to what retarding force is the effect of friction equal ?  Ans. 1–72nd of the weight.

15. Show that the space described during any interval by a body moving under the action of a constant force is the same as if it had moved uniformly during the same interval with its mean velocity.

16. The intensity of gravity at the surface of the planet Jupiter being about 2·6 times as great as it is at the surface of the earth, find approximately the time which a heavy body would occupy in falling from a height of 167 ft. to the surface of Jupiter. *U. of L. Matriculation, 1873.*

17. If a body is projected upwards with a velocity of 120 ft. in a second, what is the greatest height to which it will rise, and when will it be moving with a velocity of 40 ft. per second ? *U. of L. Matriculation, 1874.*

18. A rifle-bullet is shot vertically downwards from a balloon at the rate of 400 ft. per second. How many feet will it pass through in 2 seconds, and what will be its velocity at the end of that time, neglecting the resistance of the air ? *U. of L. Matriculation, 1875.*

19. A stone is let fall from the top of a railway carriage which is travelling at the rate of 30 miles an hour. Find what horizontal distance and what vertical distance the stone will have passed through in one-tenth of a second. *U. of L. Matriculation, 1875.*

# CHAPTER V.

## ON THE FUNDAMENTAL PROPERTIES OF FLUIDS.

92. A fluid is a body, all of whose parts can move freely amongst themselves. Motion consequently can be caused amongst the particles of a fluid body by the application of the slightest conceivable force.

93. The science which treats of the equilibrium of forces acting upon fluid bodies is termed Hydrostatics.

94. Fluids are divided into liquids and aeriform fluids. An aeriform fluid is distinguished from a liquid by the existence of an expansive or repulsive force amongst its particles, in consequence of which they tend in a greater or less degree to move off from one another.

95. When a fluid is at rest, any portion of it may be supposed to become solid, without disturbing the equilibrium.
Since all the particles of the fluid are at rest, it is clear that they will not less be so, if any number of them be rigidly connected with each other, and so deprived of the power of relative motion.

96. **Fluids transmit pressure equally and in all directions.**—In the sides of a closed vessel of any shape let a number of apertures of equal area be made, and let these apertures be supplied with pistons exactly fitting them. Let the vessel be filled with water, and let the pistons be maintained by some mechanical contrivance in their respective positions.* If any one of the pistons be pressed in with a force P, it is found that all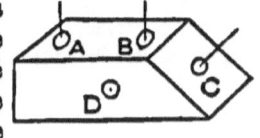

* For the reason of this provision, see Art. 105.

the other pistons experience a similar pressure, and that a corresponding force P must be applied to each of them, in order to retain them in their places. Thus the pressure communicated to the fluid has been transmitted in all directions, for all the pistons experience it, whatever their position and whatever the shape of the vessel; and the transmitted pressure is equal to that communicated.

These results are the same, whatever the number of the pistons, and wherever placed. It follows, therefore, that every portion of the surface of the containing vessel, equal in area to that of the piston, experiences a pressure P. Thus, if the area of a piston is one square inch, and a pressure of 1 lb. is exerted upon it, a corresponding pressure of 1 lb. is experienced by every square inch of the surface of the containing vessel.

97. If A and B, the areas of whose lower surfaces are $a$ and $b$ respectively, be two pistons fitted into the upper surface of a vessel filled with fluid, and if Q be the pressure experienced by B when a pressure P is exerted upon A, then

$$Q : P :: b : a.$$

For in consequence of the pressure upon the piston A, a pressure P is transmitted to every portion of the surface of the containing vessel, whose area equals $a$, and therefore to every unit of area a pressure is transmitted equal to $P \div a$.

Since the lower surface of B contains $b$ units of area, the whole pressure experienced by B is $Pb \div a$. But by the hypothesis, Q denotes this pressure, therefore $Q = Pb \div a$, that is

$$Q : P :: b : a,$$

$$\text{or} \quad \frac{\text{pressure on B}}{\text{pressure on A}} = \frac{\text{area of B}}{\text{area of A}}.$$

As an illustration, let the lower surface of the piston A be a circle whose diameter is 1 inch, and that of B a circle whose diameter is 1 foot, then

$$\frac{\text{area of B}}{\text{area of A}} = \frac{12^2}{1^2} = 144,$$

and therefore,

$$\frac{\text{pressure on B}}{\text{pressure on A}} = 144,$$

or

$$\text{pressure on B} = 144 \times \text{pressure on A};$$

so that if a weight of 1 lb. rest upon the piston A, a weight of
144 lbs. must be placed upon the piston B in order to maintain
equilibrium; and conversely, if 144 lbs. press on the piston B,
the equilibrium may be maintained by a pressure of 1 lb. on A.

98. **Bramah press.**—This powerful machine, invented by
Mr. Bramah, is an interesting application to practical purposes
of the characteristic property of fluids.

C and D are two cylindrical vessels, connected with each
other by means of the pipe *ab*.  A and B are two solid pistons

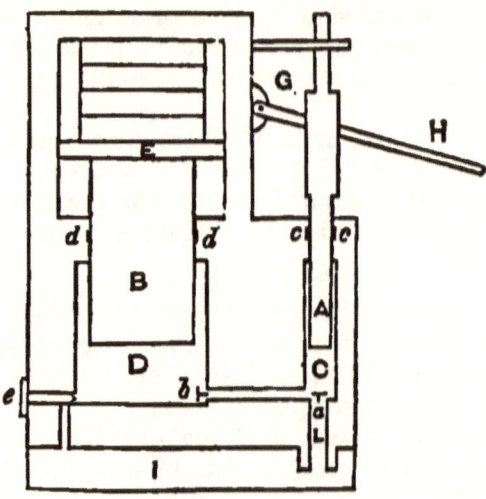

working in water-tight collars *cc*, *dd*.  The piston B, the diameter
of which is much greater than that of A, supports a plate E, upon
which the substance to be pressed is placed.  A is capable of
being worked up and down by a lever GH, having its fulcrum
at G.  L is a pipe leading into a reservoir I, *a* is a valve opening
upwards, and *b* a valve opening into the vessel D.

Let the vessels C, D be supposed to be filled up with water,
when the pistons are in the position represented in the figure.
Let A be forced down with a pressure P, then, by the preceding
article, B is forced up with a pressure $= P \times \dfrac{\text{area of B}}{\text{area of A}}$.

If B yield to this pressure, the piston A will descend, and a
portion of the water in C will be forced into the vessel D.  The
return of the fluid will be prevented by the valve *b*.  If the

piston A be now raised, fresh water will be pumped up into the vessel C from the reservoir I, and if A be again forced down with a pressure P, B will be again forced up with the same pressure as before. This process may be continued as long as the substance yields to the pressure exerted upon it.

The pressure upon B may at any time be removed by unscrewing the plug e, by which the water is allowed to flow back into the reservoir.

99. The safety valve is another important application of the same principle. A portion of the boiler of a steam-engine, whose area contains a square inches, is furnished with a valve opening outwards. The valve is so contrived as to admit of being loaded with certain weights. If a weight of aP lbs. be placed on the valve, it is pressed down with a pressure of P lbs. to the square inch. If then at any time the pressure of the steam be greater than P lbs. to the square inch, the valve will be forced open, and the steam will escape; and since the pressure of the steam is the same on every inch of the surface of the containing vessel, no portion of the boiler will experience a pressure greater than P lbs. to the square inch. So that if P be less than the pressure per square inch which the weakest portion of the boiler can bear, the boiler can never burst.

100. All the particles of a fluid, equally with those of any solid body, are subject to the action of gravity. Any portion, however small, of a fluid is drawn towards the earth with a certain degree of force—that is, is possessed of weight. (Art. 32.) The total pressure, therefore, of a fluid at rest upon any surface in contact with it may be, and most commonly is, the sum of two different pressures, the one a pressure transmitted simply by the fluid, and the other a pressure arising from the weight of the fluid itself. The former, as we have seen, is the same upon every unit of a surface in contact with the fluid. The latter, it will be seen hereafter, varies with the position of the surface. When the latter is known in any case, it is only necessary to add to it the amount of transmitted pressure corresponding to the area of the surface, and the total pressure is found. If then we can determine the pressure of a fluid upon any surface arising from the action of gravity upon its particles, upon the supposition that no external pressure is communicated to the fluid, the problem is completely solved.

**101. Pressure of fluids upon surfaces in contact with them.**—The pressure of a fluid at rest upon any *plane* must be in a *direction* perpendicular to the plane.

For the only force opposed to the pressure of the fluid on the plane is the resistance of the plane itself; and since there is equilibrium, these two forces must be equal and opposite. The resistance of the plane is perpendicular to the plane; this consequently must be the direction of the pressure of the fluid.

*The pressure of a fluid upon any plane in contact with it is equal to the weight of a column of the fluid whose base is the area of the given plane, and whose height is the depth of the centre of gravity of the plane below the surface of the fluid.*

Let A, B, C, D be moveable pistons, with plane surfaces, fitted into the sides of any vessel. If the vessel be filled with water, or any other fluid, the pressure of the fluid will force these pistons outwards. The amount of this pressure may be learnt by finding the force necessary to keep the pistons in their places. As the pressure acts at right angles to the surface of the piston, this force must also act at right angles to the piston. Let a cord be attached to any one of these pistons, as at A, and pass over pulleys so placed that the cord forms a right angle with the surface of the piston. Let W be the weight which must be attached to the other extremity of the cord in order to keep the piston in its place. It

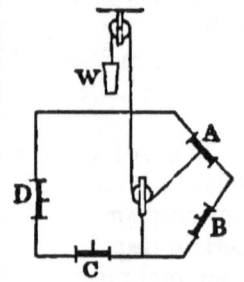

will be found that W is equal to the weight of a column of the fluid whose base is the area of the piston A, and whose height is the depth of the centre of gravity of the surface of the piston below the surface of the fluid. Thus, if the area of the piston be 6 square inches, and the depth of its centre of gravity be 8 inches, W will be equal to the weight of 6 × 8, or 48 cubic inches of the fluid. By similar experiments with other pistons such as B, C, and D, whose surfaces are differently inclined, it will be found that the result is independent of the inclination of the surface; so that if the surface be the same, and the depth of its centre of gravity be the same, the pressure of the fluid is the same, whatever the inclination of the given surface may be.

*Ex.* 1. To find the pressure on a rectangular plane 10 inches by 4, when immersed in water so that its centre of gravity is at a depth of 20 inches.

The area of the plane is 10 × 4, or 40 square inches. Hence the pressure on the plane is the weight of 40 × 20, or 800 cubic inches of water.

The weight of a cubic foot of pure water is 1000 oz. avoirdupois, and therefore the weight of a cubic inch is 1000 ÷ 1728 oz.

$$\therefore \text{ pressure required} = \frac{800 \times 1000}{1728}$$
$$= 462 \cdot 96 \text{ oz.}$$

*Ex.* 2. To compare the pressures on the bottom and side of a cubical vessel filled with any fluid.

Let $a$ be the length of the edge of the cube, and $w$ the weight of a cubic unit of the fluid. The area of the bottom of the vessel is $a^2$, and the depth of its centre of gravity is $a$, ∴ the pressure on the bottom $= wa^3$.

The area of the side of the vessel is $a^2$, and the depth of its centre of gravity is $\frac{1}{2}a$, ∴ the pressure on the side is $\frac{1}{2}wa^3$. Hence the pressure on the side of the vessel is one-half of that on the bottom.

102. What has been stated above of the pressure of a fluid upon plane surfaces applies also to curved surfaces. The *total pressure of a fluid upon any surface is equal to the weight of a column of the fluid, whose base is the area of the given surface, and whose height is the depth of the centre of gravity of the surface below the surface of the fluid.* Thus, if A be the area of any surface, $h$ the depth of its centre of gravity, and $w$ be the weight of a cubic unit of the fluid, then the total pressure of the fluid upon the surface is in all cases

$$wAh.$$

N.B.—By total pressure is meant the sum of the pressures exerted at all the points of the surface. These pressures act at right angles to the surface, and hence, when the surface is curved they are not all parallel, and, consequently, their sum is not the same as their resultant. It is only in the case of plane surfaces that the total pressure and the resultant pressure are identical. In Art. 108 we shall see how the resultant pressure may be found, when the surface is the entire surface in contact with the fluid. Other cases are too difficult to be introduced here.

*Ex.* 1. A globe, whose radius is 6 inches, is immersed in water so as to be just covered, required the total pressure upon its surface.

Here $w = \frac{1000}{1728}$ oz. ; $A = 4 \times \frac{22}{7} \times 36$, and $h = 6$

$$\text{pressure required} = \frac{1000 \times 4 \times 22 \times 36 \times 6}{1728 \times 7}$$

$$= 1571\frac{3}{7} \text{ oz.}$$

*Ex.* 2. A cone, whose height is 12 inches, and the radius of its base 9 inches, is immersed in water, with its vertex on the surface, and its axis vertical, required the total pressure on the conical surface.

The area of a conical surface is found by multiplying half the circumference of the base by the slant height. (Newth's *Mathematical Examples*, p. 77.) The slant height is $\sqrt{(9^2 + 12^2)}$ or 15. Hence, $A = \frac{198}{7} \times 15$.

The centre of gravity of a conical surface is distant from the base by one-third of the height. Hence, $h = \frac{2}{3} \times 12 = 8$. Therefore,

$$\text{pressure required} = \frac{1000 \times 198 \times 15 \times 8}{1728 \times 7}$$

$$= 1964\frac{2}{7} \text{ oz.}$$

## 103. Pressure of a fluid upon a surface independent of the quantity of the fluid.

The total pressure of a fluid upon a surface in contact with it being in all cases equal to $wAh$, is dependent only upon these three things, viz., the density of the fluid, the area of the surface, and the depth of the centre of gravity of the surface. If these three remain the same, the pressure remains the same. Hence, if a surface of any size or shape be placed in contact with a fluid mass, with its centre of gravity at a given depth, the total pressure is not affected by any alteration in the quantity of the fluid.

For instance, if the bottoms of any number of vessels of different shapes be of the same size, and the same fluid be poured into each, to the same height, the pressures upon the bottoms will be the same, notwithstanding the differences in the quantity of the fluid in the several vessels.

This may be shown experimentally in the following way.

Let A, B, C be three vessels of the forms represented in the figure, supported on a frame by projecting rims.   The bottom of each is of the same size, formed of a metal plate, moveable on a hinge, and kept close by a catch, loaded with a weight.

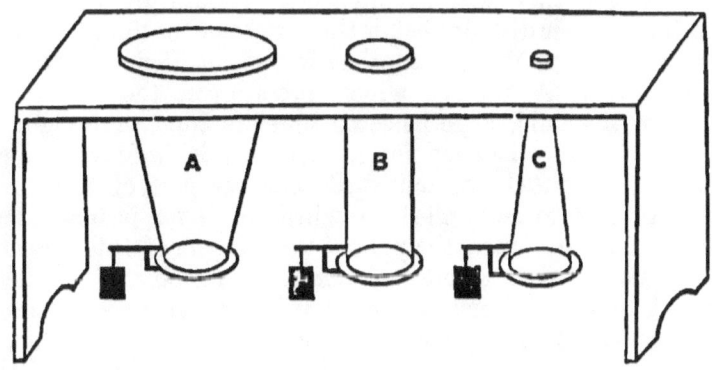

Let water be poured into the vessel C, until the pressure overcomes the weight, and forces open the bottom.   Let the same weight be attached to the vessel B, and let water be poured into this vessel, it will be found that the bottom will remain closed until the water rises to the same height as that to which it rose in C.   Repeat the experiment with A, and the same result will happen.   This shows, that when the water is at the same height in the three vessels, the pressure on the bottom is the same, notwithstanding that the quantity of water in the vessels is greatly different.

104. Another consequence from the fact that the value of the total pressure of a fluid upon a surface depends only upon the three quantities, $w$, A, and $h$, is, that if a surface be placed in contact with a given fluid, with its centre of gravity at a given depth, the amount of the fluid pressure is not affected by turning the surface about its centre of gravity.

If, for example, a circular plate of any size be immersed in water, with its centre at a depth of 1 foot, the pressure upon it is the same in *amount*, whether the surface be horizontal or vertical, or inclined at any angle whatever.   The only effect of altering the inclination of the surface is to alter the *direction* of the fluid pressure.

105. If the student will now revert to Art. 96, he will see why it was necessary to insert the provision, that the pistons be maintained in their position by some mechanical contrivance. For if the pistons (as is the case with the pistons C, D) be situated any where except upon the uppermost surface of the vessel, they will experience a pressure greater or less according to their distance below the surface of the fluid, and will, in consequence of this pressure, be forced out, if not prevented.

The pistons A and B, which press upon the uppermost surface of the fluid, experience no pressure from the weight of the fluid itself; whatever force, therefore, is impressed upon them is transmitted undiminished to every part of the fluid. But if a pressure be applied elsewhere, at C for instance, and no other force act upon the piston, then a part of this pressure will be employed in counterbalancing the pressure of the fluid upon C, and the remainder only will be the pressure transmitted by the fluid. Hence, if $a$ be the area of any piston, $h$ the distance of its centre of gravity below the surface of the fluid, and $w$ the weight of a cubic unit of the fluid, and if $a\text{P}$ be the pressure exerted upon the piston, the pressure transmitted to every corresponding area is $a\text{P} - wah$, or the pressure transmitted to every unit of area is $\text{P} - wh$.

106. *If a fluid be poured into any one of a number of open vessels having a free communication with each other, the fluid will rise to the same height in all.*

Let A and B be any two such vessels, having the lowest point at C.

Let a thin vertical plate of the fluid passing through C become rigid. Let $a$ be the area of this plate, $h$ the distance of its centre of gravity below the surface of the fluid in A, and $h'$ its distance below the surface in B.

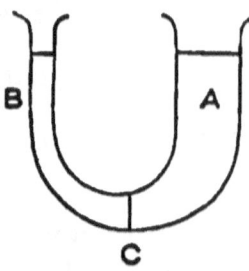

Let $w$ be the weight of a cubic unit of the fluid. Then the pressure exerted on the one side of the plate by the fluid in A is $wah$, and that exerted on the other side by the fluid in B is $wah'$. But since there is equilibrium, these pressures must be equal; therefore

$$wah = wah',$$
or
$$h = h'.$$

**107. Hydrostatic bellows.**—AB is a narrow tube communicating with a vessel formed by uniting together two pieces of wood by some flexible and waterproof substance. If water be poured down the tube, it will enter the vessel, and raise a large weight W placed upon CD.

When equilibrium exists, let A be the surface of the water in the tube, and let B lie in the same horizontal plane with CD. Let $w$ be the weight of a cubic unit of water, and let $h$ be the depth of B or CD below A.

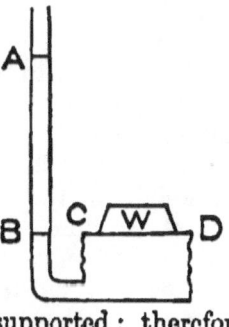

The pressure upon each unit of area in CD is $wh$. Therefore, if A = number of units in the area of CD, the total pressure on $CD = wAh$, and since there is equilibrium, this must equal the weight supported; therefore

$$W = wAh.$$

If $a$ = area of the horizontal section of the pipe at B,

weight of fluid in $AB = wah$,

∴ $$W = \frac{A}{a} \times \text{weight of fluid in AB.}$$

**108. The resultant pressure of a fluid on the surface of a solid, wholly or partially immersed in it.**

Let any portion M of any fluid at rest become solid, and since the equilibrium is not thereby disturbed, the forces acting upon M must be in equilibrium. The only forces acting upon M are the weight of M acting vertically downwards at the centre of gravity of M, and the pressures of the fluid upon the surface of M. Hence the resultant of these pressures must be equal and opposite to the weight of M acting at its centre of gravity. But the fluid will exert the same pressure upon the surface of any other body of the same form as M

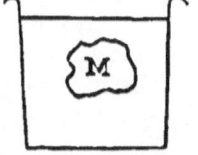

occupying its place. Hence, *the resultant pressure of a fluid on the surface of a solid wholly or partially immersed in it, is equal to the weight of the fluid displaced, and acts vertically upwards through the centre of gravity of the fluid displaced.*

Hence also if a body be immersed partly in one and partly in the other of two fluids that do not mix, the resultant pressure of the fluids is equal to the weight of the displaced fluids.

**109. Bodies weighed in fluids.**—If any body hang freely by a cord, we have seen (Art. 37) that the centre of gravity lies in the vertical line drawn through the point of suspension, that is, in the line of the cord; and the tension in the cord equals the weight of the body.

If such a body be wholly immersed in a fluid, it has been shown in the preceding article, that it will be pressed upwards by a force equal to the weight of the fluid displaced, which, in this case, is the weight of a portion of the fluid equal to the bulk of the body; and this pressure is directly opposite to the weight of the body; therefore the tension in the cord will be the weight of the body diminished by the weight of an equal bulk of the fluid.

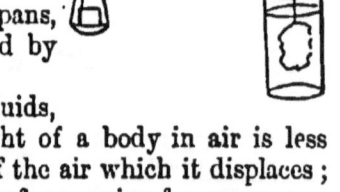

Hence, when a body is weighed in a fluid, as may be done by means of the Hydrostatic balance, which in its simplest form is only a common pair of scales, with a fine thread or wire attached to the under surface of one of the scale pans, the weight of the body is diminished by the weight of the fluid displaced.

This statement applies to all fluids, whether liquids or gases. The weight of a body in air is less than its true weight by the weight of the air which it displaces; the loss of weight being at the rate of 31 grains for every 100 cubic inches in its volume.

**110. Conditions of equilibrium of a floating body.**— If a body floating in a fluid be at rest, the forces acting upon it are the weight of the body, which acts at its centre of gravity, and the resultant pressure of the fluid, which acts at the centre of gravity of the fluid displaced.

The necessary conditions of equilibrium are, that these forces be *equal* and *opposite*. The two conditions, therefore, are,

First, that the weight of the body be equal to that of the fluid displaced; and,

Secondly, that the centres of gravity of the body and the fluid displaced be in the same vertical line.

As a body cannot displace a quantity of fluid greater than its own bulk, if the weight of the body be greater than that of an equal bulk of the fluid, the body cannot float in that fluid. If the weight of the body be exactly equal to that of an equal bulk of the fluid, the body will have in no position any tendency either to rise or to sink, provided only it be entirely covered by the fluid.

## EXAMPLES.

1. Two circular pistons, whose diameters are 2 and 10 inches respectively, are fitted into the upper surface of a vessel filled with fluid; if a pressure of 10 lbs. be applied to the smaller piston, what is the pressure on the larger piston?                                    Ans. 250 lbs.

2. The diameter of the large piston in a Bramah press is 20 inches, of the small $\frac{1}{4}$ inch, the lever is 2 feet long, and is attached to the small piston rod at a point 2 inches from the fulcrum, what is the mechanical advantage of the press?                              Ans. 76800.

3. Find the pressure upon a rectangular plane 12 in. by 5, when immersed in water, so that its centre of gravity is at a depth of 24 inches.
                                                            Ans. 833$\frac{1}{3}$ oz.

4. The diameter of the plate of a hydrostatic bellows is 12 inches, a weight of 250 lbs. is placed upon it, what will be the height of the water in the pipe?                                 Ans. 61$\frac{1}{11}$ in. nearly.

5. A cubic foot of fir weighs 40 lbs., to what depth will it be immersed when floating in water?                          Ans. 7·68 in.

6. If a cubic foot of oak, when floating in water, be immersed to the depth of 9 inches, what is its weight?         Ans. 46 lbs. 14 oz.

7. A globe, 6 inches in radius, when floating in water is half immersed, what is its weight?                             Ans. 261$\frac{11}{14}$ oz.

8. A conical vessel, with its vertex downwards, is filled with water, required the total pressure upon its surface, the height being 24 in., and the radius of its base 7 in.                     Ans. 2546$\frac{8}{17}$ oz.

9. A cylindrical piece of wood is immersed in water, so that its centre is at the depth of 12 inches, required the total pressure on the cylindrical surface, the length of the cylinder being 14 in., and the radius of its base 3 in.                                    Ans. 1833$\frac{1}{3}$ oz.

10. A vessel in the shape of a pyramid 5 feet high, and with a base 4 feet square, is filled with water; find the pressure on the base. *U. of L. Matriculation, 1870.*

11. Find the pressure on a vertical rectangle, 10 inches long and 6 inches broad, immersed in water with its longer sides horizontal, and with the upper one 2 inches below the surface. *U. of L. Matriculation, 1871.*

12. Explain under what circumstances the pressure of a liquid on the bottom of a vessel containing it is different from the force with which it presses against the table on which it rests. Account for this difference of pressure. *U. of L. Matriculation, 1871.*

# CHAPTER VI.

ON SPECIFIC GRAVITY.

111. The specific gravity of any substance is the ratio of the weight of the substance to the weight of an equal bulk or volume of a certain standard substance.　For solids and liquids the standard is distilled water at a temperature of 60° Fahrenheit.　For gases the standard is atmospheric air at a temperature of 60°, with the barometer at 30 inches.

Hence, for solid and liquid bodies,

$$\text{sp. gr.} = \frac{\text{weight of body}}{\text{weight of equal volume of water}};$$

and for gaseous bodies,

$$\text{sp. gr.} = \frac{\text{weight of body}}{\text{weight of equal volume of air}}.$$

It follows from this definition, that the weight of any body is equal to its specific gravity multiplied by the weight of an equal volume of water (or air).　Hence, if W be the weight of any body, V its volume, $s$ its specific gravity, and $w$ the weight of a unit of the standard, then

$$W = V s w.$$

A cubic foot of pure distilled water at 60° weighs 1000 oz. avoirdupois, and 100 cubic inches of air at the standard temperature and pressure weigh 31 grains.　Hence, for solids and liquids, if a foot be taken as the unit of measurement, $w = 1000$ oz.; but if an inch be the unit, $w = \frac{1000}{1728}$ oz.　For gases, taking an inch as the unit, $w = \cdot 31$ grains.*

---

* Using the French system of measures, a litre (or cubic decimetre) of water at its greatest density is 1000 grammes, and a litre of air at 0° C, and with the barometer at 760 millimetres, is 1·293 grammes.

From these data any one of the following three quantities, viz., the weight, the volume, and the specific gravity, can be found when the other two are known.

*Ex.* 1. To find the weight of a cubical block of marble, whose side is 8 feet.

The volume of the block is 8 × 8 × 8, or 512 cubic feet. The sp. gr. of marble is 2·7. Therefore,

$$\text{weight of block} = 2\text{·}7 \times 512 \times 1000 \text{ oz.}$$
$$= 38\text{·}571 \text{ tons.}$$

*Ex.* 2. To find the weight of a cubic foot of chlorine gas. The volume in inches is 1728.

The sp. gr. of chlorine is 2·5. Therefore,

$$\text{weight req.} = 2\text{·}5 \times 1728 \times \text{·}31 \text{ gr.}$$
$$= 3\text{·}061 \text{ oz. av. nearly.}$$

*Ex.* 3. A body weighs 3 lbs., and its specific gravity is 7·2, required the volume.

The weight of the body in ounces is 3 × 16, or 48. Hence,

$$48 = V \times 7\text{·}2 \times \tfrac{1000}{1728}.$$
$$\therefore \quad V = \frac{48 \times 1728}{7200} = 11\text{·}52 \text{ in.}$$

*Ex.* 4. What volume of oxygen gas will weigh 1 lb. av.?

The weight of the gas in grains is 7000, the sp. gr. is 1·11. Hence,

$$7000 = V \times 1\text{·}11 \times \text{·}31$$
$$\therefore \quad V = \frac{7000}{\text{·}3441} = 20342\text{·}923 \text{ in.}$$
$$= 11\text{·}772 \text{ feet.}$$

**112. Methods of finding the specific gravity of a solid body.**—*First: To find the specific gravity of a body heavier than water.*

Weigh the body both in air and in water. The loss of weight in water is, as seen in Art. 109, equal to the weight of the water displaced; that is, since the body is entirely immersed, of a volume of water exactly equal to that of the given body. Therefore,

$$\text{sp. gr.} = \frac{\text{weight of the body}}{\text{loss of weight in water}}.$$

*Secondly: To find the specific gravity of a body lighter than water.*

To the given body attach some other body heavy enough to sink it in water; weigh the two together both in air and in water; the loss of weight is equal to the weight of the water displaced by both. Then weigh the heavy body in air and in water; the loss of weight is equal to the weight of the water displaced by the heavy body.

The difference of the two losses is therefore equal to the weight of water displaced by the given body. Hence,

$$\text{sp. gr.} = \frac{\text{weight of body}}{\text{difference of the two losses}}.$$

The specific gravity of a body lighter than water may also be found by allowing it to float in water, and then measuring the volume of the part immersed. For if V be the volume of any body, and $s$ its specific gravity, then its weight (Art. 111) is $Vsw$. And if V′ be the volume of the part immersed, that is, of the water displaced, then the weight of the water displaced is V′$w$. But when a body floats, its weight is equal to the weight of the displaced fluid. Therefore,

$$Vsw = V'w,$$

$$\therefore \quad s = \frac{V'}{V} = \frac{\text{volume displaced}}{\text{whole volume}}.$$

If the body be cylindrical, and float with its axis vertical, then the proportion of the volume displaced to the whole volume is the same as the proportion of the depth immersed to the whole depth. Hence, in such cases,

$$\text{sp. gr.} = \frac{\text{depth immersed}}{\text{whole depth}}.$$

*Thirdly: To find the specific gravity of a compound when the weights and specific gravities of the components are known.*

Let $W_1 W_2$, &c., be the weights, $V_1 V_2$, &c., the volumes, and $s_1 s_2$, &c., be the specific gravities of the components. Let W be the weight of compound, V the volume, and $s$ the required specific gravity. Then if no change of volume result from the composition,

$$V = V_1 + V_2 + \&c.$$

But by Art. 111, $V = \dfrac{W}{sw}$; $V_1 = \dfrac{W_1}{s_1 w}$, and so on. Therefore

$$\frac{W}{sw} = \frac{W_1}{s_1 w} + \frac{W_2}{s_2 w} + \&c.$$

$$\therefore \quad \frac{W}{s} = \frac{W_1}{s_1} + \frac{W_2}{s_2} + \&c.$$

From which equation $s$ may be found, since all the other quantities are known.

If the volume be increased by the mixture, the specific gravity is proportionately diminished. If, on the other hand, the volume be diminished by the mixture, the specific gravity is proportionately increased. Thus, for example, if the volume of the compound is $\frac{5}{4}$ of the sum of the volumes of the components, the sp. gr. is $\frac{4}{5}$ of the value determined above; if the volume of the compound is only $\frac{3}{4}$ of the sum of the volumes of the components, the sp. gr. is $\frac{4}{3}$ of this value.

*Fourthly: To find the specific gravity of a compound when its volume and the volumes and specific gravities of the components are known.*

Using the same notation as before,

$$W = W_1 + W_2 + \&c.$$

But, Art. 111, $W = Vsw$; $W_1 = V_1 s_1 w$, and so on. Therefore

$$Vsw = V_1 s_1 w + V_2 s_2 w + \&c.$$

$$\therefore \quad Vs = V_1 s_1 + V_2 s_2 + \&c.$$

Whence $s$ may be found, all the other quantities being known.

113. If in the first and second methods described in the preceding Art. some other liquid had been used instead of water, we should have obtained the specific gravity of the solid body relatively to that of the liquid; and hence, if a body be weighed in any liquid, and its loss of weight be observed,

$$\frac{\text{sp. gr. of body}}{\text{sp. gr. of liquid}} = \frac{\text{weight of the body}}{\text{loss of weight in the liquid}};$$

and so if a body float in any liquid,

$$\frac{\text{sp. gr. of body}}{\text{sp. gr. of liquid}} = \frac{\text{volume displaced}}{\text{whole volume}}.$$

These results may also be deduced as follows: Let W be the weight, V the volume, and $s$ the specific gravity of the body. Let $l$ be its loss of weight in any liquid whose specific gravity is $s'$, then Art. 111 ($w$ being the weight of a unit of water),

$$V_{s}w = W$$

and, Art. 109.          $$V_{s'}w = l$$

$$\therefore \quad \frac{s}{s'} = \frac{W}{l}$$

Again, let V' be the volume displaced when the body floats in the liquid, then, the weight of the floating body being equal to that of the displaced liquid,

$$V_{s}w = V's'w$$

$$\therefore \quad \frac{s}{s'} = \frac{V'}{V}.$$

## 114. Methods of finding the specific gravity of a liquid.

*Firstly : To determine the specific gravity of a liquid by means of the specific gravity bottle.*

The specific gravity bottle is simply a small flask fitted with a ground stopper.

Let $w$ be the weight of the flask, $x$ its weight when filled with the given liquid, and $y$ its weight when filled with distilled water. Then $x - w =$ the weight of the given liquid contained in the flask, and $y - w =$ the weight of a similar volume of distilled water.

$$\therefore \quad \text{sp. gr. of the fluid} = \frac{x - w}{y - w}.$$

Specific gravity bottles are very frequently made so as to contain exactly 1000 grains of pure distilled water. Then, if $x$ and $w$ are known in grains,

$$\text{sp. gr. of the liquid} = \frac{x - w}{1000}.$$

*Secondly : To determine the specific gravity of a liquid by weighing a solid body in it.*

Weigh the body in air, in the given liquid, and in water. The loss of weight in the given liquid is equal to the weight of the liquid displaced ; and the loss of weight in water is equal to the weight of water displaced. The volumes of the displaced liquid and of the displaced water are the same. Therefore,

$$\text{sp. gr.} = \frac{\text{loss of weight in given liquid}}{\text{loss of weight in water}}.$$

*Thirdly: To determine the specific gravity of a liquid by means of a hydrometer.*

The COMMON HYDROMETER consists of a small hollow sphere A, to which there is attached on one side a slender graduated stem, and on the other a smaller sphere B, made of such a weight as to allow the whole instrument to float with the sphere A entirely immersed.

A table accompanies the instrument, in which are given the specific gravities corresponding to the several divisions of the stem. The specific gravity of any liquid is found by observing the degree to which the instrument sinks when placed in it, and then referring to the tables.

NICHOLSON'S HYDROMETER consists of a hollow sphere, or some other symmetrical figure, connected to a stem which carries a cup at each end. The lower cup is loaded, so as to secure stability of equilibrium when the instrument floats with the stem vertical.

When in use, the instrument is sunk to an invariable depth, by means of a weight placed in the upper cup; the volume of liquid displaced is therefore always the same.

Let W be the weight of the instrument, and $w$ the weight which must be placed in B to sink the instrument to the point D in distilled water; then the weight of water displaced is equal to

$$W + w.$$

Let $w'$ be the weight required to sink the instrument to D when placed in any given liquid, then the weight of the liquid displaced is equal to

$$W + w'.$$

The volumes displaced in each case are the same, therefore

$$\text{sp. gr. of the liquid} = \frac{W + w'}{W + w}.$$

## 115. Method of finding the specific gravity of a gas.

—The method of finding the specific gravity of a gas is similar to that of finding the specific gravity of a liquid by means of the specific gravity bottle. Instead of the bottle a flask is used, fitted with a stop-cock, and capable of being screwed on to the end of an exhausting syringe, or to the plate of an air-pump.

The air is first exhausted from the flask, and the stop-cock being closed, the flask is carefully weighed. We thus obtain the weight of the empty flask. The flask is then weighed when filled with air and with the gas. Deducting from these weights the weight of the empty flask, we obtain the weight of the air and of the gas contained in the flask, and the volumes being equal,

$$\text{sp. gr.} = \frac{\text{weight of gas in the flask}}{\text{weight of air in the flask}}.$$

As the volume of the flask will commonly differ greatly from that of the weight by which it is balanced, the flask and the weight will be differently affected by the buoyancy of the surrounding air; and although this will not sensibly affect the accuracy of the result if the atmospheric pressure remain the same throughout the entire series of operations, it is otherwise if any change in the atmospheric pressure takes place. Hence, when great accuracy is required, provision must be made to guard against this source of error. This is conveniently done by attaching to the scale-pan (b) containing the weight, a closed flask (B) whose exterior volume is equal to that of the flask (A) containing the gas. Both arms of the balance are thus equally affected by the atmospheric pressure under all circumstances, and the difference between the weight in (b) when the flask (A) is exhausted, and the weight when the flask is filled with any gas will give the weight of the contained gas. It is almost needless to state that in such experiments a very sensitive balance should be used.

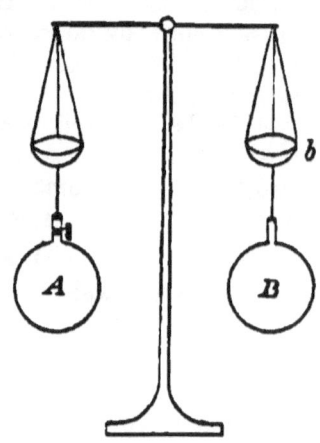

## TABLE OF SPECIFIC GRAVITIES.

| | | | |
|---|---|---|---|
| Platinum (forged) | 21·4 | Beech | ·85 |
| Gold | 19·3 | Oak (English) | ·75 |
| Mercury (congealed) | 15·6 | Scotch Fir | ·69 |
| —— (fluent) | 13·6 | Spruce Fir | ·52 |
| Lead | 11·3 | Cork | ·24 |
| Silver | 10·5 | Sulphuric Acid | 1·8 |
| Copper | 8·9 | Nitric Acid | 1·5 |
| Brass 7·8 to | 8·4 | Honey | 1·45 |
| Steel | 7·8 | Blood (human) | 1·05 |
| Iron (forged) | 7·7 | Ale (average) | 1·04 |
| Tin | 7·3 | Milk | 1·03 |
| Iron cast at Carron | 7·2 | Sea Water | 1·03 |
| Zinc (compressed) | 7·2 | Wine (Port) | ·99 |
| —— (common) | 6·9 | Castor Oil | ·97 |
| Aluminium | 2·6 | Proof Spirit | ·93 |
| Sodium | ·97 | Olive Oil | ·92 |
| Potassium | ·87 | Turpentine | ·87 |
| Diamond | 3·5 | Brandy | ·84 |
| Plate Glass | 2·9 | Alcohol | ·80 |
| Marble | 2·7 | Ether Sulphuric | ·72 |
| Cornish Granite | 2·7 | Chlorine Gas | 2·50 |
| Green Glass | 2·6 | Carbonic Acid | 1·53 |
| Flint (black) | 2·6 | Nitrous Oxide | 1·53 |
| Portland Stone | 2·5 | Oxygen | 1·11 |
| Brick | 2·0 | Nitric Oxide | 1·04 |
| Ivory | 1·8 | Carburetted Hydrogen | ·97 |
| Sugar | 1·6 | Nitrogen | ·97 |
| Lignum Vitæ | 1·3 | Phosphuretted Hydrogen | ·90 |
| Box (Dutch) | 1·3 | Ammoniacal Gas | ·59 |
| Ebony | 1·2 | Sub-Carburetted Hydrogen | ·56 |
| Mahogany ·64 to | 1·1 | Hydrogen | ·07 |

### EXAMPLES.

1. What is the weight of a block of gold 8 in. long, 6 in. wide, and 4 in. deep?   Ans. 134 lbs. of oz. avoirdupois.
2. What is the weight of a cast-iron globe whose radius is 12 in. ?   Ans. 16 cwt. 93 lbs. 11¾ oz.
3. A body weighs 340 grains in air, and 102 grains in water, what is its specific gravity ?   Ans. 1·429.
4. A lump of iron weighing 2 lbs. 4 oz. in air, and 1 lb. 15 oz. in water, is attached to a piece of cork, the two together weigh 2 lbs. 10 oz. in air and 12 oz. in water, what is the sp. gr. of the cork ?   Ans. ·24.
5. If 12 lbs. 1 oz. of gold be mixed with 11 lbs. 2 oz. of copper, what is the sp. gr. of the compound ?   Ans. 12·37.
6. If 20 cubic inches of tin be mixed with 30 inches of copper, what is the sp. gr. of the compound ?   Ans. 8·26.

7. A cube, whose edge measures 12 inches, sinks in water to a depth of 8 inches, but sinks in a given fluid to the depth of 10 inches, what is the sp. gr. of the fluid ?                                           Ans. ·8.

8. A body, whose weight is 300 grains, sinks to the same depth in water and spirits of wine, when loaded with 80 and 17 grains respectively, what is the sp. gr. of the spirits of wine ?                Ans. ·834.

9. A cylindrical piece of cork, whose length is 20 inches, floats upright in water, to what depth will it be immersed ?                   Ans. 4·8 in.

10. To what depth will the same piece of cork be immersed in a fluid whose sp. gr. is ·72 ?                                             Ans. 6⅔ in.

11. The sp. gr. of a mixture of brandy and water is ·936, how much brandy is there in a pint of the mixture ?                          Ans. ⅔ pt.

12. A body when floating in a liquid, whose sp. gr. is ·84, is immersed to the extent of two-thirds of its volume, what is its sp. gr. ?    Ans. ·56.

13. A cylindrical body, whose length is 12 in. and sp. gr. ·8, floats upright in two liquids which do not mix, whose specific gravities are ·92 and ·74, to what depth will it be immersed in the heavier fluid ?
                                                                      Ans. 4 in.

14. A piece of metal weighs 211·6 grains in vacuo, 187·32 grains in water, and 182·37 grains in a solution of sodic chloride. Find the sp. gr. of the solution. *U. of L. Matriculation, 1869.*

15. A piece of cupric sulphate weighs 3 oz. in vacuo, and 1·86 oz. in oil of turpentine. What is the sp. gr. of the cupric sulphate, that of turpentine being ·88 ? *U. of L. Matriculation, 1868.*

16. When equal volumes of alcohol (sp. gr. = ·8) and distilled water are mixed together, the volume of the mixture (after it has returned to the original temperature) is found to fall short of the sum of the volumes of its constituents by 4 per cent. Find the sp. gr. of the mixture. *U. of L. Matriculation, 1872.*                                          Ans. ·9375.

17. Taking 7·67 lbs. as the weight of 100 cubic feet of air, find approximately the volume of hydrogen (sp. gr. compared with air = ·07) which a balloon must contain in order that its total lifting power may be equal to the weight of 713 lbs. *U. of L. Matriculation, 1872.*
                                                                Ans. 10000 cub. ft.

18. An inch cube of ice (sp. gr. = ·918) is floating in water of which the sp. gr. is ·99987. Find accurately what will be its height above the surface of the water. *U. of L. Matriculation, 1875.*

19. An accurate balance is totally immersed in a vessel of water. In one scale pan some glass (sp. gr. 2·5) is being weighed, and exactly balances a one pound weight (sp. gr. 8·0) which is placed in the other scale plane. Find the real weight of the glass. *U. of L. Matriculation, 1875.*                                                Ans. 1 lb. 7⅓ oz.

# CHAPTER VIL

## ON ATMOSPHERIC PRESSURE.

116. The atmosphere surrounding the earth, being a fluid acted on by gravity, presses upon all bodies immersed in it, in accordance with the general laws of fluid pressure.

The pressure of the atmosphere upon any body will consequently vary with the depth of that body below its highest surface. As, however, the dimensions of any ordinary body upon the surface of the earth are inconsiderable, when compared with the height of the atmosphere, this pressure may, without any sensible error, be regarded as the same at all points, in any such body.

It is found by experiment that the pressure which the atmosphere exerts upon a body, on or near the surface of the earth, is at the rate of about 15 lbs. for every square inch of surface. That we are not, in ordinary circumstances, cognizant of this pressure arises from the characteristic property of fluids, in accordance with which they press in all directions upon bodies immersed in them; so that, if the upper surface of a body, subject to the pressure of the atmosphere, experience a downward pressure of 15 lbs. to every square inch, the under surface experiences a corresponding upward pressure. If, however, we destroy this equilibrium, by removing or diminishing the pressure of the atmosphere upon one side of any body, we are immediately made sensible of the pressure which is exerted upon the opposite side. For instance, if a plate of glass be placed upon the top of a cylindrical receiver connected with an air pump, and the air be exhausted from the receiver, the plate of glass will be pressed down with a considerable force, and if not strong enough will be broken in pieces.

117. **Atmospheric pressure variable.**—The atmospheric pressure at the same place is not found to be constant, but

undergoes considerable variation. Roughly speaking, it is found
to vary between $13\frac{3}{4}$ lbs. and $15\frac{1}{4}$ lbs. per square inch. It has
been found, however, to be more convenient to measure the
atmospheric pressure not by pounds or ounces, but by the length
of a column of a fluid, whose weight would be equal to the
atmospheric pressure upon its base. Mercury is the fluid most
commonly employed for this purpose, and the column of mercury
whose weight is equal to the atmospheric pressure is found to
vary between 28 and 31 inches.

Mercury is 13·6 times as heavy as water, and therefore the
column of water which is equal to the atmospheric pressure will
range between 32 ft. and 35 ft. The average is a little under
30 inches of mercury, or 34 feet of water.

The instruments by which the variations of the atmospheric
pressure are observed and measured are termed **Barometers.**

118. *When the atmosphere presses upon any point of a fluid
mass, to determine the amount of transmitted pressure.*

If the atmosphere presses upon the highest surface of the
fluid mass, then it is evident that the transmitted pressure is
equal to the pressure of the atmosphere.

If the atmosphere presses at a point in the fluid mass below
the highest surface, then a portion of the atmospheric pressure
will be employed in counterbalancing the outward pressure of
the fluid itself, and the remaining portion only of the atmospheric pressure will be transmitted by the fluid. Let $h$ be the
distance below the highest surface of the fluid mass of the point
at which the atmosphere presses upon it, and let $w$ be the weight
of a cubic unit of the fluid, then, as seen in Chapter V., the
fluid itself will exert at this point an outward pressure on every
square unit $= wh$. Then, if $Æ$ denote the pressure exerted by
the atmosphere upon a square unit, the transmitted pressure
will be
$$Æ - wh$$
for every unit of surface in the fluid mass.

If the point at which the atmosphere presses upon the fluid
mass be so taken, that $wh$ is greater than $Æ$, then it is plain
that equilibrium cannot subsist, but motion will ensue amongst
the particles of the fluid, until the distance becomes such that
$wh = Æ$.

119. **The cistern barometer.**—This instrument is constructed in the following manner:—A glass tube, closed at one

end, is filled with mercury, and, a finger being placed over the open end, is inverted in an open vessel of mercury.

Let AB be such a tube, let CD be the surface of the mercury in the vessel, and let MB be more than 31 inches. Then, since CD is more than 31 inches below B, the highest surface of the fluid mass, the pressure of the fluid upon any square unit in the surface CD will be greater than the atmospheric pressure. The mercury will consequently fall in the tube. Let N be the point at which it rests, then if Æ be the atmospheric pressure, and $w$ the weight of a cubic unit of mercury,

$$Æ = w \cdot MN.$$

The height of mercury in the tube will accordingly vary as the atmospheric pressure varies; will rise as the atmospheric pressure increases, and fall as it decreases.

A graduated scale is attached to the tube AB, by means of which the changes in the height of the mercury in the tube can be observed and measured.

120. It will be seen, that when the mercury falls in the tube it rises in the cistern, and that consequently the fall of the mercury in the tube, as measured by the scale attached to it, will not correctly measure the variation in the barometrical column, but will be too great by the distance through which the mercury has *risen* in the cistern. In like manner, when the mercury rises in the tube, the variation, as measured by the scale, will be too small by the distance through which the mercury has *fallen* in the cistern.

If the cistern be large in comparison with the bore of the tube, the variation in the height of the mercury in the cistern is so small that it may be disregarded for ordinary purposes. When greater accuracy is desired, the bottom of the cistern is so made that it can be raised or lowered by a screw, and thereby the surface of the mercury in the cistern be adjusted to one uniform height.

121. **The siphon barometer.**—In this instrument the necessity of any adjustment is avoided by the simple device of substituting for the cistern a small tube, of the same bore as that of the larger tube. Let ABC be such an instrument, the surfaces of the mercury being at A and C. Suppose now the

mercury to fall in the shorter tube from C to c, and in consequence to rise in the longer tube from A to a. Then,
since the bore of both tubes is alike, the distance
Aa is equal to Cc. The entire variation in the
barometric column is Aa + Cc, and therefore equal
to 2 Aa. Hence the variation, as measured by the
scale, is exactly one-half of the real variation.

This form of the instrument possesses some
advantages, but is open to the objection that by
diminishing the scale it increases the difficulty of
observing small variations of the height of the
column.

122. **The wheel barometer.**—This form of the
barometer has been contrived for the purpose of
making small variations of the column more sensible. It consists of a siphon barometer, having a
small ball of iron or glass floating upon the open
surface of the mercury. A fine thread, attached to
the ball, passes round a small wheel, carrying an index, which
moves over a graduated dial plate, and the rise or fall of the
mercury is measured by the arc of the circle over
which the index moves. The variations of the
barometric column range between 28 and 31 inches,
or through a distance of 3 inches. As shewn in
the last section, the variation at C is one-half of
the variation in the barometric column; and, consequently, the ball will rise or fall through a space
of 1½ inch. If, then, the circumference of the
wheel round which the ring passes be exactly 1½
inch, the index will make a complete revolution,
when the ball rises or falls through a height of 1½
inch; and if the circumference of the dial plate
be divided into 360 equal parts, the motion of the
index over 120 of these parts will correspond to a
rise or fall of the ball through half an inch, and
therefore to a change of one inch in the barometric
column. Consequently, the variation of one 120th
part of an inch is shown by the motion of the index over one
division of the plate.

123. **The siphon.**—The siphon is a bent tube, open at both

ends, and having one leg shorter than the other. If such a tube be filled with any fluid, and placed with the extremity of its shorter leg in a vessel containing the same fluid, the fluid will flow out through the longer leg.

Let ABC be the siphon filled with fluid, and having B, the extremity of its shorter leg AB, placed below MN, the surface of the fluid in the vessel. Let A be the highest point of the siphon, and through A draw the vertical line AF. Let MN produced meet this vertical line in D, and let CF be the horizontal line drawn through C.

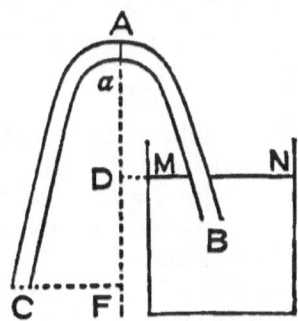

Let Aa be a thin vertical film of the fluid passing through A. Since each side of this film is similarly situated with respect to A, the highest point of the fluid mass, the pressure arising from the weight of the fluid will be the same on each side. These pressures, therefore, are in equilibrium, and do not cause the motion of the fluid.

But if Æ be the atmospheric pressure upon a square unit, and $w$ be the weight of a cubic unit of the fluid, then, since the atmosphere presses upon the surface MN at a distance AD below the highest point of the fluid mass, there is transmitted up the leg BA, by Art. 118, a pressure upon every square unit equal to $Æ - w . AD$.

Similarly, since the atmosphere presses upon the surface of the fluid mass at C, there is transmitted up CA a pressure upon every square unit equal to $Æ - w . AF$.

Hence, the one side of the film Aa will experience a pressure upon every square unit equal to $Æ - w . AD$, and the other side a pressure equal to $Æ - w . AF$. But since AD is less than AF, $Æ - w . AD$ is greater than $Æ - w . AF$; the film of fluid will therefore be forced down the leg AC. The same will happen to every film that may in succession occupy the position Aa; the fluid consequently will flow out at C, so long as there remains in the vessel any fluid above B.

It has been assumed in the preceding that the height AD is such that $w . AD$ is less than Æ. If AD be so great that $w . AD$ is greater than Æ, then, as seen in the preceding articles, the fluid will sink in the leg AB to some level below the bend, and the siphon consequently will cease to act.

The conditions therefore that a siphon may act for any fluid are, first, that the distance of the highest point from the surface of the fluid is less than the length of a barometric column of the fluid; and, secondly, that the open end of the siphon is below the surface of the fluid. If these conditions be fulfilled, it is indifferent whether the legs of the siphon are equal or unequal; and if unequal, whether the longer or shorter leg be inserted in the fluid.

. 124. **The common pump.**—The common pump consists of a cylindrical barrel AB, furnished with a valve at B opening upwards, and with an air-tight piston, capable of being moved up and down by means of a rod.

The piston is also furnished with a valve opening upwards, and to the barrel at B is attached a pipe BC, called the suction pipe.

Let C be the surface of the water to be raised, and suppose the piston to be at B.

Let the piston be raised from B to A; the air in BC will, by virtue of its expansive force, press open the valve at B and fill the barrel AB; and, in conse-quence of occupying a larger space ABC, will exert a less pressure than before upon the surface of the water within the pipe. Hence, the pressure of the air within the pump being less than the atmospheric pressure upon the surface of the water, the water will be forced up the pipe BC until equilibrium be restored.

Let the piston now descend; the valve at B will be closed, and the air in AB will escape through the valve in D. If the piston be again raised, the preceding process will be repeated, and the water will again rise in the pipe BC. In like manner, upon every successive stroke of the piston the water will con-tinue to rise in the pipe, until, if BC be less than the water barometric column, it at last reaches B.

Let the water be at B, and let the piston be pressed down to B. Then, since the atmosphere presses upon the surface of the fluid at C, there will be transmitted by the fluid a pressure upon every square unit of the valve at B equal to $Æ - w$ . BC (Art. 118); and, consequently, when the piston is again raised, the water will force open the valve at B, and will rise in the barrel as far as A, if AC be less than the water barometric column, or to some point between B and A, if AC be greater than the barometric column.

Upon the next descent of the piston, the valve in D will be forced open, and the water will flow through it; so that when the piston reaches B the water in the barrel will lie above the piston, and when the piston is raised will be raised along with it, and flow out at the spout A.

It will be seen, that if BC be greater than the water barometric column, no water will rise into the barrel; and in such a case the pump will be useless. In order therefore that the pump may work in all states of the atmosphere, BC must be less than the lower limit of the water barometric column, that is, must be less than 32 feet. If BC be less, but AC more, than the water barometric column, the water will only in part fill the barrel, and the discharge of water during the ascent of the piston will not be continuous.

125. **The forcing pump.**—The construction of the forcing pump is represented in the accompanying figure.

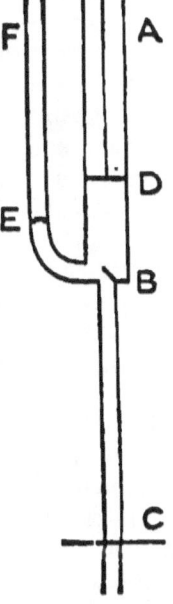

D is a piston with no valve, working in a barrel AB; BC is a pipe descending to the water; EF a tube leading upwards from B; at B and E are valves opening upwards.

If the piston be raised from B to A, then, as in the common pump, the air in BC will press open the valve at B and fill the barrel, and the water will rise to a certain height in BC. As the piston descends, the air in the barrel will be forced out through the valve at E; after a certain number of ascents the water will, if BC be less than 32 feet, pass into the barrel AB; and, upon its next descent, the piston will force the water in the barrel through the valve E into the pipe EF. As the piston continues to rise, more water will pass into the barrel, and this in its turn will be forced, by the descent of the piston, into the pipe EF.

126. **The double-barrel air pump.**—AB and CD are two cylindrical barrels connected together by the pipe AC, and

communicating by means of another pipe with an air-tight
receiver R. M and N are pistons furnished with valves opening
upwards, and are worked by the toothed wheel O, so that when
M descends N will ascend, and *vice versâ.* At A and C are
valves opening upwards.

Let M be at A, and N at D, and the wheel be turned so that
N may descend and M ascend. As N descends the valve at C
will close, and the air in the barrel CD will pass out through
the valve in N. As M ascends, the pressure of the external
air will close the valve in M ; and the pressure being removed
from the upper side of the
valve A, the air in the receiver will in consequence of
its expansive force press open
the valve and fill the barrel
AB. The air originally in the
receiver now occupying a larger
space, viz., the receiver and
barrel together, will be of
diminished density.

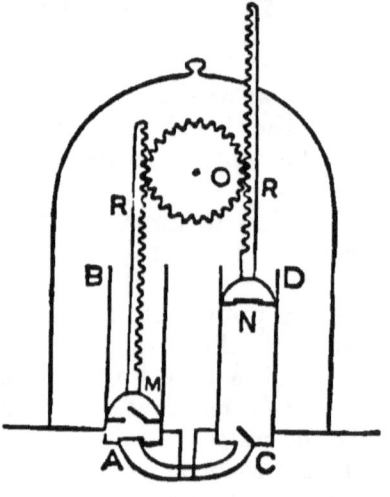

If the wheel be again turned
a similar process will take place,
and the air in the receiver will
again expand, so as to fill both
receiver and barrel, and a
second diminution of density
will consequently take place.

The process may be continued so long as the air in the
receiver can by its expansive force press open the valves at A
or C; but since the expansive force of the air is diminished
with the diminution of its density, the action of the pump
must, after a certain number of strokes, entirely cease.

127. **The single-barrel air pump.**—This air pump consists of a single barrel AB, communicating with an air-tight
receiver by the pipe C.

The piston passes through an air-tight collar in the plate BD.
At B is a valve opening upwards.

The process of exhaustion is precisely similar to that described
in Art. 126.

The effect of the valve at B is to relieve the piston from the
pressure of the atmosphere during a part of its ascent. For if

the air in BM be of less density than atmospheric air, the pressure of the atmosphere will press down the valve at B, and

will keep it closed until the piston by its ascent has so compressed the air in BM, that its density has become greater than that of atmospheric air.

The plate BD with its valve opening upwards is not peculiar to the single-barrel air pump, but may, if required, be adopted in a double-barrel air pump. In fact, it is used in the more carefully constructed double-barrel pumps.

**128. The Sprengel air pump.—** By means of this instrument a much nearer approach to a perfect vacuum can be attained than is possible with the common air pump. It consists of a long glass tube, BCDE, bent as in the figure, and having the limb CD longer than the barometric column. The bore of the longer arm DE and of the upper part of CD is very fine, about one-tenth of an inch in diameter. At B the tube is connected by means of a piece of canvas-covered india-rubber tubing with the pipe of the funnel A. Connection between the funnel and the tube CD can be closed by means of a clamp at B. At D is a short branch tube connected by the india-rubber tubing F with the receiver R.

When the instrument is in action the funnel A is filled with mercury, and the clamp at B being loosened the mercury rises in CD, and falls down DE in successive drops. As the mercury passes the opening of the branch tube at D it carries with it a portion of the air from the receiver, and DE becomes filled with a number of columns of mercury separated by small columns of air. As DE is longer than the barometric column for mercury, the mercury, and with it the enclosed air, will run out at E into a vessel placed there to receive it.

As the exhaustion proceeds the number of columns of air become fewer and fewer, until at length the lower part of the tube DE shews a continuous column of mercury nearly equal in height to the barometric column. When this is so the exhaustion is complete, no more globules of air being enclosed within the mercury. The time required for reaching this point varies of course with the size of the receiver. If the receiver be about a pint in volume from 20 to 30 minutes is required. The instrument itself gives notice to the operator of the approaching completion of the operation by the sharp metallic noise which result when a liquid falls in a vacuum.

By means of the Sprengel pump the density of the air in the receiver can be reduced to less than one millionth part of that of the atmosphere.

129. **The condensing air pump.**—This differs from the single-barrel air pump only in having the two valves opening towards the receiver instead of away from it. As the piston descends the valve in the piston becomes closed, and the valve at the bottom of the barrel opens, and the air in the barrel is forced into the receiver. When the piston is raised, the elastic force of the air in the receiver closes the valve in the barrel, and so prevents the egress of the contained air. The pressure of the atmosphere opens the valve in the piston, and the barrel is thus a second time filled with air.

It will be seen that by each descent of the piston a quantity of air is forced into the receiver equal, in its original volume, to the content of the barrel.

By a simple arrangement, the same barrel may be made available as both an exhausting and a condensing syringe. The piston C is solid. Two tubes A and B furnished with valves lead from the bottom of the barrel, the valves both opening in the same direction (as

represented in the figure both open to the left). If the tube
A be connected with the receiver the
syringe will act as a condensing syringe,
the air entering at B during the *upward*
stroke of the piston. If the tube B be
connected with the receiver, the syringe
will act as an exhausting syringe, the
air being expelled through A during
the *downward* stroke of the piston.

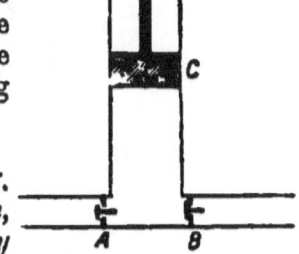

### 130. Boyle and Mariotte's law.

*The temperature remaining the same,
the pressure (or elastic force) of any
gas varies inversely as the space it occupies,*

The experiments by which this law is established are of the
following kind.

A glass tube of uniform bore, open at both ends, is bent into
the form ABC; the shorter branch, BC, is furnished with a
stop-cock C, and an accurately divided scale is attached to the
longer branch.

Let the stop-cock be opened, and a small quantity of mercury
be poured into the tube. Let the surfaces of the mercury be at
D and E; these will be in the same level, since
the same pressure, viz., the atmospheric pressure,
acts upon both. Let the stop-cock now be closed,
the surfaces D and E will remain at the same
level, and the pressure of the air in CE, upon the
surface at E, is equal to the pressure of the atmosphere upon D; or the elastic force of the air in
CE is equal to the atmospheric pressure. Let the
barometer stand at *h* inches, and let *w* be the
weight of a cubic inch of the mercury used in
the barometer, then the atmospheric pressure per
inch, and consequently the elastic force of the air
in CE, is equal to

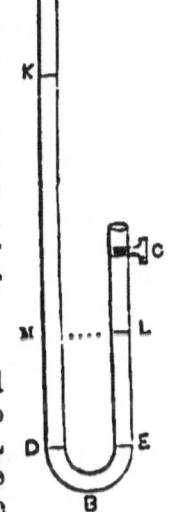

$$wh$$

Now let mercury of the same temperature and
density as that in the barometer be poured into
the larger branch, and let the mercury rise to L
in the shorter branch, and stand at K in the
longer branch; then the pressure at L must be
equal to the pressure at M. The pressure at M per inch is the

atmospheric pressure, increased by the weight of KM inches of mercury; hence, the elastic force of the air in CL is equal to

$$wh + w \,.\, KM$$

Hence we obtain this result,

$$\frac{\text{elastic force of the air in CL}}{\text{elastic force of the air in CE}} = \frac{wh + wKM}{wh},$$

$$= \frac{h + KM}{h}.$$

It is found, by trial, that

$$\frac{CE}{CL} = \frac{h + KM}{h};$$

and, therefore, it follows that

$$\frac{\text{elastic force of air in CL}}{\text{elastic force of air in CE}} = \frac{CE}{CL},$$

or the elastic force of the air varies inversely as the space it occupies.

By similar experiments the law is established for other gases.

Hence, if $p_1$ be the elastic force, and $V_1$ the volume of a given mass of gas at a given temperature, and if $p_2$ be the elastic force when the same mass has become compressed or expanded into the volume $V_2$, the temperature remaining the same; then

$$p_1 : p_2 :: V_2 : V_1.$$

In performing the experiments described above, care must be taken to guard against any increase of temperature by the sudden compression of the gas. The mercury should be poured in very slowly, and some time allowed before observing the levels, that any heat which may have been generated may pass away by radiation.

Since the density of any body varies inversely as the space it occupies, it follows that Boyle and Marriotte's law may be thus enunciated. *The temperature remaining the same, the elastic force of any gas varies directly as the density.*

131. The following are examples in illustration of Boyle's law :—

*Ex.* 1. Determine the elastic force of a mass of air, whose volume is 100 cubic inches, when compressed into 40 cubic inches; the barometer standing at 30 inches.

The elastic force of the air, before compression, is equal to the atmospheric pressure, that is, to the weight of 30 inches of mercury, or 14·757 lbs. Hence, if $p$ be the elastic force required,

$$p = \frac{100 \times 14·757}{40},$$
$$= 36·892 \text{ lbs.}$$

*Ex.* 2. The contents of the receiver and barrel of an air pump are as 5 to 2, compare the pressure of the air after 3 strokes of the piston with that of the atmosphere.

Let $p$ be the atmospheric pressure, and $p_1$, $p_2$, $p_3$ the pressures of the contained air after the first, second, and third strokes respectively.

After any stroke the air that was in the receiver fills both receiver and barrel, that is, its volume changes from 5 to 7. Hence,

$p_1 : p :: 5 : 7$    or    $p_1 = \tfrac{5}{7}p$

Also    $p_2 : p_1 :: 5 : 7$    ,,    $p_2 = \tfrac{5}{7}p_1 = \left(\tfrac{5}{7}\right)^2 p.$

and    $p_3 : p_2 :: 5 : 7$    ,,    $p_3 = \tfrac{5}{7}p_2 = \left(\tfrac{5}{7}\right)^3 p.$

Atmospheric pressure is commonly measured by the column of mercury which it sustains. Hence, if the mercurial barometer stand at $h$ inches, the pressure of the air per inch after 3 strokes will be equal to the weight of $\left(\tfrac{5}{7}\right)^3 \times h$ inches of mercury.

*Ex.* 3. The content of the receiver of a condensing air pump is 40 cubic inches, and that of the barrel is 5 cubic inches, by how much will the pressure of the air in the receiver be increased after 10 strokes of the piston?

By each stroke of the piston 5 cubic inches of air are forced into the receiver, and hence after 10 strokes 50 cubic inches of air are forced in. Hence, with the air originally in the receiver 90 cubic inches of air have now the volume of 40 cubic inches. Hence,

$$\frac{\text{pressure of enclosed air}}{\text{atmospheric pressure}} = \frac{90}{40} = \frac{9}{4}.$$

## EXAMPLES.

1. What is the atmospheric pressure per inch when the barometer stands at 30 inches?      Ans. 14 lbs. 12 oz.

2. What change in the atmospheric pressure will be denoted by a change of 1 inch in the height of the barometric column? Ans. $7\tfrac{1}{3}$ oz.

3. If a barometer be formed of a liquid whose sp. gr. is 7·5, at what height will it stand when a quicksilver barometer stands at 30 inches ?

Ans. 54·4 in.

4. Will the siphon act if the shorter leg be inserted into a closed vessel ?

5. What is the greatest available length of the shorter leg of a siphon used for drawing off water, the barometer standing at 30 inches ?

Ans. 34 ft.

6. What is the greatest available length of the shorter leg of a siphon used for drawing off a liquid whose sp. gr. is 2·5 ; the barometer standing at 29½ inches ?                                         Ans. 13·37 ft.

7. If the spout of a common pump be 15 feet above the surface of the water to be raised, and the area of the piston be 10 square inches, what power will be required to work the pump, the handle being a lever whose longer arm is 30 inches, and shorter 3 inches ?          Ans. 6 lbs. 8⅓ oz.

8. If the spout of a common pump used for raising a liquid whose sp. gr. is 1·2 be 14 feet above the surface of the liquid, and the area of the piston be 8 square inches, what power will be required to work the pump, the handle being a lever whose longer arm is 3 feet and shorter 3 inches ?                                         Ans. 4 lbs. 13⅞ oz.

9. Find the sp. gr. of the heaviest liquid that can be drawn off by a siphon whose shortest leg is 54·4 inches, the barometer standing at 30 inches.                                         Ans. 7·5.

10. If the content of the receiver of an air pump be 3 times that of the barrel, and if the barometer stand at 768 millimetres, by what column of mercury will the elastic force of the air be measured after 4 strokes of the piston ?                                         Ans. 243.

11. An air-tight piston fits into a closed cylindrical vessel, shew that the pressure necessary to force it in three-fourths of the length is equal to that of three atmospheres.

12. An air-tight piston fitted into a closed cylindrical vessel is placed at a given distance from the bottom of the vessel, shew that the pressure necessary to draw it out to twice that distance is equal to that of half an atmosphere.

13. A barometer placed in a cylindrical diving bell, and standing at 30 inches when starting, is observed to stand at 32 in., how high will the water have risen in the diving bell, the height of which is 6 feet ?

Ans. 4½ in.

14. In the preceding, at what height will the barometer stand when the water has risen 15 inches ?          Ans. 37·9 in. nearly.

15. A U tube, 1 foot in height, and having one limb closed and filled with mercury, is placed under an air pump, what ratio does the elasticity of the air bear to that of the atmosphere when the mercury has fallen through 4 inches, the barometer standing at 30 inches.   Ans. As. 2 : 15.

16. A cylindrical bell 4 ft. deep, whose content is 20 cubic feet, is lowered into water until its top is 14 ft. below the surface of the water, and air is forced into it until it is three-quarters full, what volume would the air occupy under the atmospheric pressure, the water-barometer being at 34 ft. ?  U. of L. Matriculation, 1874.

17. 40 cubic centimetres of atmospheric air are enclosed in a tube over mercury, the height AB of the mercury in the tube above the level in the trough being 50 centimetres.  The tube is depressed until AB is equal to

30 centimetres.  What is now the volume of the air [the height of the barometer (76 centimetres) and the temperature have not changed during the observations].  *U. of L. Matriculation, 1875.*

Ans. 22$\frac{11}{23}$ cubic centimetres.

18.  The capacity of the barrel of a condensing air pump is 10 cubic inches, and of a copper receiver 100 cubic inches.  By how much will the pressure of the air in the receiver be increased after 20 strokes of the piston?  *U. of L. Matriculation, 1875.*

19.  A tumbler of air is placed mouth downwards under water at such a depth that the surface of the water inside it is at the depth of 25½ ft. Compare the weight of a cubic inch of air in the tumbler with that of a cubic inch of the air outside,—the barometer standing at 30 inches and the sp. gr. of mercury being 13·6.  *U. of L. Matriculation, 1871.*

Ans. As 7 : 4.

20.  A receiver attached to an air-pump has the volume of 100 cubic inches, while the cylinder has the volume of 10 cubic inches, what proportion of the original air will be left in the receiver after the completion of the fourth double stroke?  *U. of L. Examination for Women, 1874.*

21.  At the bottom of a mine a mercurial barometer stands at 77·4 centimetres; what would be the height of an oil barometer at the same place, the specific gravity of mercury being 13·596 and that of oil ·9. *U. of L. Matriculation, 1876.*

22.  The two limbs of a Mariotte's tube are graduated in inches.  The mercury in the shorter limb stands at the graduation 4, and five inches of air are enclosed above it.  The mercury in the other limb stands at the graduation 38.  The barometer at the time indicates a pressure of 29·5 inches.  Find to what pressure the five inches of air are subjected, and also the length of tube they would occupy under the atmospheric pressure alone.  *U. of L. Matriculation, 1876.*

23.  A certain quantity of air at atmospheric pressure has a volume of 2 cubic feet, the temperature being 55° Fahr.  What does the volume of the air become when the pressure is increased by one-twentieth, the temperature meanwhile remaining the same?  *U. of L. Examination for Women, 1876.*

# CHAPTER VIII.

## ON LIGHT.

132. *Light* is the name by which we describe the cause which produces the sensation of vision. Respecting its nature and mode of action nothing is certainly known. Different theories have been proposed with the view of explaining the various phenomena of light, but success in these attempts has hitherto been but partial.

133. A *self luminous* body is one which by itself excites the sensation of vision. Bodies not self-luminous may excite the sensation of vision, but can do so only when some self-luminous body is present; such bodies simply transmit light which has originated elsewhere.

134. **Medium. Transparency.**—Any portion of space through which light is propagated is called a *medium*, whether the space be occupied by any body or not. Bodies through which light can be propagated are called *transparent;* those which entirely intercept it are called *opaque.* Strictly speaking, these terms do not distinguish different bodies from each other, but express qualities which, to a greater or less degree, belong alike to all bodies. There is no known substance so transparent as not to intercept some portion of light, nor, on the other hand, is any substance so opaque as not to transmit some light.

135. **Propagation of light in straight lines.**—Experience teaches us that light, unless acted upon by some external cause, is propagated in straight lines. If, for instance, we take a hollow tube, and there be present no cause capable of changing the direction of light, we find that we can see through the tube only when straight lines can be drawn along it. If the tube be so bent that no straight line can be drawn along it, we cannot see through it.

**136. Ray. Pencil. Focus.**—A *ray* of light is the smallest conceivable portion of light, and is represented by the straight line along which it is supposed to be propagated. A collection of rays is called a *pencil* of light. A parallel pencil is one whose component rays are all parallel to each other, a converging pencil one in which the rays all proceed *to* a single point, and a diverging pencil one in which they all proceed *from* a single point. The point to or from which the rays proceed is called the *focus of the pencil.*

**137. Reflection and refraction.**—When light falls upon the surface of any body, one portion, greater or less, is turned back into the medium whence it came, or, as it is termed, *reflected;* another portion, greater or less, is transmitted through the new medium, and undergoes a change in its direction, termed *refraction;* while a third portion, greater or less, is absorbed by the new medium.

**138. Angle of incidence. Angle of reflection. Angle of refraction.**—When a ray of light falls upon any surface, the angle which it makes with the perpendicular to the surface at the point of incidence is termed the *angle of incidence;* the angle which the reflected ray makes with the same perpendicular is termed the *angle of reflection;* and, similarly, the angle which the refracted ray makes with this perpendicular is termed the *angle of refraction.*

N.B. The perpendicular to the surface is sometimes called the *normal.*

**139. Laws of reflection.**—*First: The reflected ray lies in the same plane with the incident ray and the normal at the point of incidence. Secondly: The angle of reflection is equal to the angle of incidence.*

Thus, if AO be a ray of light incident upon the reflecting surface MN, and if OB be the reflected ray, and OP the normal, then will OB lie in the same plane with AO and OP, and the angle BOP be equal to the angle AOP.

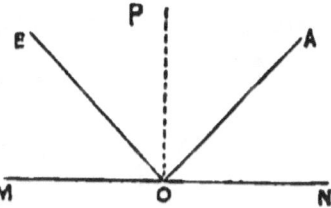

I

140. *If a diverging pencil of light fall upon a plane reflect-ing surface, the focus of the reflected pencil will be at the same distance from the surface as that of the incident pencil, but on the opposite side of it.*

Let P be the focus of the incident pencil, and AC a section of the reflecting surface. Draw PN perpendicular to AC, and make $pN = PN$. Let PA be any inci-dent ray, join $pA$, then shall $pA$ produced, or AD, be the direction of the reflected ray. For, by the construction, the triangles $pNA$, PNA are equal, and therefore the angle $ApN$ = the angle APN. But if AM be the perpendicular to the surface at A, then AM and P$p$

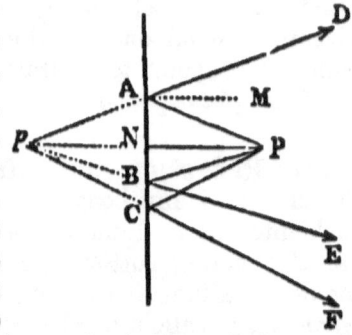

being parallel, the angle DAM = the angle $ApN$, and the angle PAM = the angle APN. Therefore the angle DAM = the angle PAM, or AD is the direction of the reflected ray. In a similar manner it may be shewn, that all the reflected rays will proceed from $p$, and therefore $p$ is the focus of the reflected pencil.

141. **Images.**—From every point in a luminous body, whe-ther self-luminous or illuminated, light is propagated in all possible directions. If, then, light from such a body fall upon any reflecting surface, or be transmitted through any refracting medium, each point may be regarded as the focus of an incident pencil. For every incident pencil there will then be a reflected or refracted pencil, each with its own focus. This set of points, made up of the foci of the reflected or refracted pencils, is termed the *image* of the luminous body. When the reflected or refracted rays actually pass through these points, the image is called *real;* but when they do not so pass, the image is called *virtual.*

142. **Position and magnitude of the image formed by a plane mirror.**

Let PQ be the given object, and AB a section of the reflect-ing surface. Draw P$p$ perpendicular to AB, and make $pA$ = PA. Then (Art. 140) $p$ is the focus of the reflected pencil, for all the light falling upon the surface from P. Similarly, if Q$q$ be perpendicular to AB, and $q$B be equal to QB, $q$ will be the focus of the reflected pencil for all the light incident from Q. In like manner, for every incident pencil proceeding from points lying between P and Q, the focus of the reflected pencil will lie between $p$ and $q$. Hence, the image $pq$ will be of the same size and shape as the object PQ. The image in this case is virtual, and is situated behind the surface in such a way that each point in it is at the same distance from the surface as the corresponding point in the object.

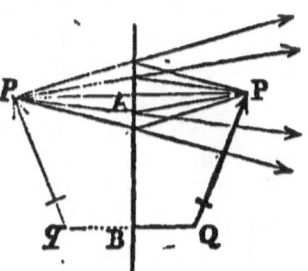

143. **Limits of the visibility of an image formed by a plane mirror.**—In many respects there is no difference, as to the effect upon vision, between an optical image and a luminous object occupying the same position. There is, however, this difference, that whereas a luminous object sends off light in every direction, an image sends off light in some directions only, limited by the magnitude of the mirror. In fact, the image is seen or not seen by an eye in front of the mirror, when a luminous object of the same size and shape, and occupying the same position, would be seen or not seen if the mirror were an opening in an opaque screen.

Let $pq$ be the image of PQ formed by the mirror MN. Draw $pM$ and $qN$ and produce to A and B. Also draw $qM$ and $pN$ and produce to C and D. Then, if MN were an opening in the opaque screen XY, and $pq$ a luminous object, $pq$ would be wholly visible to an observer stationed anywhere between the lines MA and NB; would be invisible to one beyond MC or ND, and would be partly visible to one between MA and MC, or between NB and ND. These,

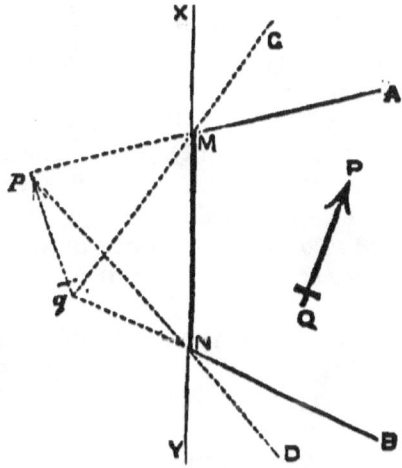

then, are the limits within which the image $pq$ is either wholly or partially visible.

**144. Path of the visual pencils, when an object is viewed in a plane mirror.**—From each point in any visible object a small pencil of light is received by the eye. The size of this pencil is limited by the size of the pupil, or the opening in the eye through which light enters. In like manner, when the eye views the image formed by a plane mirror, a small pencil enters the eye for each point in the image.

Thus, let $pq$ be the image of the object PQ formed by the plane mirror MN. Let EF be the pupil of the eye of the observer. Draw $p$E and $p$F inter-secting the mirror in A and B, then AE and BF will be the extreme rays of the pencil which en-ters the eye for the point $p$. But $p$ is the image of P, and therefore the

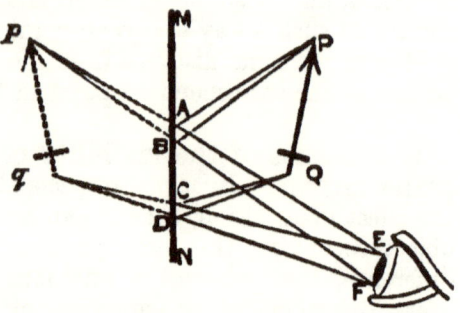

rays AE and BF came originally from P. Hence, drawing PA and PB, we obtain the complete path of the pencil which pro-ceeds from P to the mirror, and thence to the eye, producing in the mind of the observer the impression of a luminous spot at $p$. In like manner, drawing $q$E and $q$F intersecting the mirror in C and D, and QC, QD, we have the path of the visual pencil from Q to the eye, or that by which the point $q$ is seen. In a similar way the visual pencil for any other point may be obtained.

It will be seen, that for an eye in the position represented in the figure, the only part of the mirror concerned in producing vision is the part AD. The rest of the mirror might be re-moved without affecting the clearness of the vision. It will be seen also, that, for an eye in a different position, the part of the mirror concerned in producing vision will be different.

**145. Continued reflections at plane mirrors. The endless gallery.**—If an object be placed between two mirrors, the image formed by one mirror may be regarded as an object to the other mirror; and if so situated that the light reflected from the first mirror falls upon the second, then a second image will be formed by the second mirror. Under like circumstances, this image will prove an object to the first mirror, a third image be formed, and so on, as long as the light from the image formed by one mirror falls upon the other mirror.

When the two mirrors are parallel, this process may be continued without any limit, and it gives rise to what is termed the endless gallery.

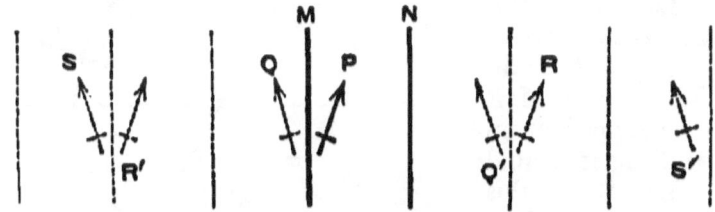

Thus, let M and N be two mirrors whose planes are parallel, and P any object placed between them. Let Q be the image of P formed by the mirror M; this will be as far behind M as P is in front of it. Since light from Q can fall upon the mirror N, an image of Q will be formed as far behind N as Q is in front of it. Let R be this image; then, since light from R can fall upon the mirror M, an image of R will be formed as far behind M as R is in front of it. Let S be this image, and then, proceeding as before, we shall obtain an endless succession of images.

Again, let Q′ be the image of P formed by the mirror N; then, proceeding as before, we obtain another series of images, Q′, R′, S′, and so on.

N.B. It is convenient to mark off on each side of the mirrors a series of distances, each equal to the distance between the mirrors. It will be found that one image will fall in each of the divisions so obtained.

146. **The Kaleidoscope.**—When the two mirrors are not parallel, the number of images of any object placed between the mirrors is limited, and is dependent upon the angle at which the mirrors are inclined: the smaller the angle the greater is the number of images. If the angle between the mirrors is an aliquot part of 360°, say the $n^{th}$ part, then the number of images is $n-1$. Thus, for example, let M and N be the two plane mirrors, inclined at an angle of 60°, or 1 − 6th part of 360°. Let the mirrors meet in O. Through O draw OA, OB, OC, OD, so that the angles MOA, AOB, BOC, COD shall each be equal to 60°, then DON also is equal to 60°.

Let Q be the image of P formed by the mirror M.   Since Q
is in front of the mirror N, an image of Q
will be formed at an
equal distance behind
N.   Let R be this
image; then, since R
is in front of the mirror M, an image of R
will be formed at an
equal distance behind
M.   Let S be this
image; but as S is
not in front of the
mirror N, no image
of S will be formed.

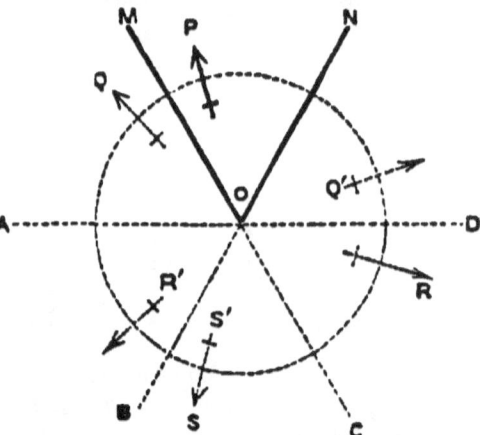

Again, let Q' be the image of P formed by the mirror N;
then, as Q' is in front of the mirror M, an image of Q' will be
formed at an equal distance behind M.   Let R' be this image;
then, since R' is in front of the mirror N, an image of R' will
be formed at an equal distance behind N.   Let S' be this
image; but S' is behind the mirror M, and therefore no image
of S' will be formed.

It will be seen, from an examination of the figure, that the
image S' must coincide exactly in position with the image S,
and hence will form but one image.   The total number of images
is in this case five.   In like manner it may be seen, that if the
mirrors had been inclined at an angle of 45°, or 1 − 8th of 360°,
the total number of images would have been 8 − 1, or 7.

### 147. Small Pencils. Direct Pencils. Axes of Pencils.—

A diverging pencil of light falling upon a plane reflecting or
refracting surface is said to be *small*, when the angle between
its extreme rays is small.   A pencil of light, whether parallel
or converging, falling upon a spherical reflecting or refracting
surface, is said to be *small*, when the angle between its extreme
rays is small; and also the area of a section of the pencil at
the surface is small, when compared with the area of the entire
sphere of which the surface forms a part.   A pencil is said to
be *direct*, when one of its rays is a normal to the surface upon
which it falls.   This ray is termed the *axis* of the pencil.

**148. Geometrical focus.**—When the reflecting surface is plane, all the rays of the reflected pencil, as already seen, diverge accurately from a single point. When the reflecting surface is spherical, the reflected rays do not in any case all proceed either to or from a single point. ' If, however, the incident pencil be small, the rays of the reflected pencil converge to, or diverge from a point so nearly, that their distance from it is inappreciable, and this point is regarded as a focus, and is called the *geometrical focus* of the reflected pencil.

**149. Principal focus, and focal length of mirror.**—If a small pencil of *parallel* rays fall directly upon a spherical mirror, the focus of the re-

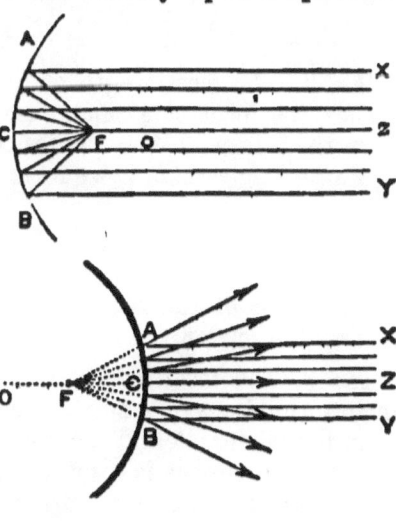

flected pencil is called a *principal focus* of the mirror. Thus, let XA, ZC, YB, be a parallel pencil falling upon the concave or convex spherical mirror AB. Let F be the focus of the reflected pencil; F is a principal focus of the mirror. Let ZC, the ray which passes through O, the centre of the spherical surface, be the axis of the pencil. The point F will lie in the axis at a distance midway between O and the mirror. The distance of F from the mirror is called the *focal length* of the mirror. Hence, *the focal length of any spherical mirror is equal to half the radius of the surface.*

It will be seen from the figures, that if F be the focus of an incident pencil, the reflected pencil will be parallel.

150. *To find the focus of the reflected pencil, when a small diverging pencil is incident directly upon a spherical mirror.*

Let $u$ = the distance from the mirror of the focus of the incident pencil, $v$ = the distance from the mirror of the focus of the reflected pencil, and $f$ = the focal length of the mirror. Then the focus of the reflected pencil lies in the axis of the incident pencil, at a point whose distance from the mirror is determined by the following formula:

$$\frac{1}{v} = \frac{1}{f} - \frac{1}{u},$$

in using which, distances measured on the side of the mirror opposite to that on which the incident light falls are to be reckoned negative.

*Ex.* 1. The focus of a small pencil of light falling directly upon a concave spherical reflector, whose focal length is 5 inches, is at a point 12 inches from the mirror, to find the distance from the mirror of the focus of the reflected pencil.

Here $u = 12$, and $f = 5$; also $f$ is positive, being measured on the same side of the mirror as the incident light,

$$\therefore \qquad \frac{1}{v} = \frac{1}{5} - \frac{1}{12} = \frac{7}{60},$$

$$\therefore \qquad v = \frac{60}{7} = 8\tfrac{4}{7}.$$

*Ex.* 2. The focal length of a concave spherical reflector is 6 inches, the focus of a small pencil incident directly is 5 inches from the mirror, to find the focus of the reflected pencil.

Proceeding as before, we have

$$\frac{1}{v} = \frac{1}{6} - \frac{1}{5} = -\frac{1}{30},$$

$$\therefore \qquad v = -30;$$

or the focus of the reflected pencil is at a distance 30 inches behind the mirror.

*Ex.* 3. The focal length of a convex spherical reflector is 8 inches, the focus of a small pencil incident directly is 4 inches from the mirror, to find the focus of the reflected pencil.

Since the surface is convex, the focal length is not measured on the same side of the mirror as the incident light, and consequently must be taken negatively; for $f$ therefore we must write $-8$. Hence,

$$\frac{1}{v} = -\frac{1}{8} - \frac{1}{4} = -\frac{3}{8},$$

$$\therefore \qquad v = -\frac{8}{3} = -2\tfrac{2}{3};$$

or the focus of the reflected pencil will be $2\tfrac{2}{3}$ inches behind the mirror.

151. From the formula just given are obtained the following general results:

When the mirror is concave (or $f$ positive), if $u$ be greater than $2f$, $v$ is positive, and between $2f$ and $f$; if $u$ be between $2f$ and $f$, $v$ is positive, and is greater than $2f$; and if $u$ be less than $f$, $v$ is negative, and greater than $u$. That is to say, if a small diverging pencil fall directly upon a concave mirror from a point beyond the centre, it will be reflected *to* a point between the centre and a principal focus; if it proceed from a point between the centre and a principal focus, it will be reflected *to* a point beyond the centre; and if it proceed from a point between a principal focus and the mirror, it will be reflected *from* a more distant point behind the mirror. In the first and second cases the reflected pencil is converging; in the third case it is diverging, but less diverging than the incident pencil.

When the mirror is convex (or *f* negative) *v* is always negative, and less than *u*. That is to say, if a small diverging pencil fall directly upon a convex mirror, it is reflected from a less distant point behind the mirror. The reflected pencil is therefore diverging, and more diverging than the incident pencil.

Hence the general effect of a concave mirror upon a diverging pencil is to diminish its divergency, of a convex mirror to increase it.

## 152. Images formed by spherical mirrors when the pencils of light are small and direct.

For greater simplicity we shall suppose the object to be either the arc of a circle concentric with the mirror, or so small as not sensibly to differ from such an arc. In such a case all the points in the object are equally distant from the mirror, and therefore the foci of all the reflected pencils will be also equally distant from the mirror.

First: Let the mirror be concave, and the distance of the object be greater than $2f$. By the preceding article the distance of the focus of each reflected pencil will be between $f$ and $2f$.

Let AB be a section of the mirror, C the centre of the spherical surface, and PQ the object. Draw the line PC; this will be the axis of the pencil which proceeds from P. The focus of the reflected pencil will be in this line, at a distance from the mirror greater than $f$, but less than $2f$; let

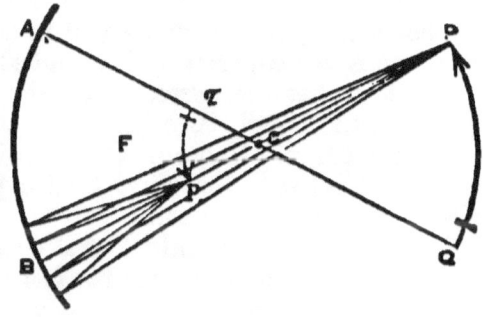

$p$ be this focus. Again draw QC, and take $q$A, equal to $p$B, then the small pencil proceeding from Q will be reflected to $q$. In like manner, for all pencils proceeding from points in the object between P and Q, the foci of the reflected pencils will lie between $p$ and $q$; $pq$ therefore is the image of PQ. In this case the image is real, inverted, and diminished.

Secondly: Let the mirror be concave, and the distance of the object be less than $2f$, but greater than $f$. Let $pq$ in the preceding figure be the object, then it is plain that PQ will be the image. Hence, in this case, the image is real, inverted, and magnified.

Thirdly: Let the mirror be concave, and the distance of the object less than $f$. Then, according to the preceding article,

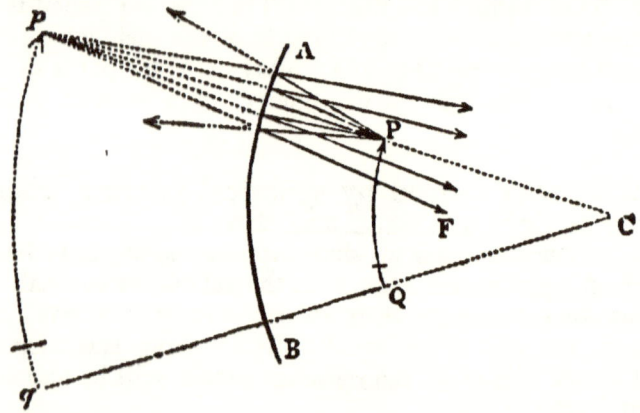

each incident pencil will be reflected *from* a more distant point behind the mirror. Hence, as before, draw the line PC, the axis of the pencil, from P. The focus of the reflected pencil will lie in PC produced, say, at $p$. In like manner the pencil proceeding from Q will be reflected from $q$, and the image $pq$ will be in the position represented in the figure. In this case the image is virtual, erect, and magnified.

Fourthly: Let the mirror be convex. If in the last figure $pq$ denote the object, then the mirror AB is convex, and PQ will be the image. Hence in all cases with a convex mirror the image falls between the mirror and a principal focus, and is erect, virtual, and diminished.

Hence with **concave mirrors**, if a small object be placed at a great distance the image is at the focus.

As the object moves from a great distance to the centre, the image moves from the focus to the centre.

As the object moves from the centre to the focus, the image moves from the centre to infinity.

As the object moves from the focus to the mirror, the image moves from an infinite distance behind the mirror to the mirror.

With **convex mirrors,** if a small object be placed at a great distance from the mirror, the image is at the focus; and as the object moves on to the mirror the image moves from the focus to the mirror.

**153. The relative magnitudes of the image and object are as their distances from the mirror.**—For the image and object being similar arcs of circles are as the radii, or as $pC$ to $PC$; that is, are as $2f - v : u - 2f$. But by Art. 150

$$\frac{1}{v} = \frac{1}{f} - \frac{1}{u}$$

$\therefore$
$$\frac{1}{v} - \frac{1}{2f} = \frac{1}{2f} - \frac{1}{u}$$

$\therefore$
$$\frac{2f - v}{2fv} = \frac{u - 2f}{2fu}$$

$\therefore \quad 2f - v : u - 2f :: 2fv : 2fu :: v : u.$

Hence $\qquad$ image : object :: $v : u$.

*Ex.* 1. The focal length of a concave spherical mirror is 5 inches, to find the nature, position, and magnitude of the image of a small object placed 30 inches from the mirror.

The object being small, each point is supposed to lie at the same distance from the mirror, namely, 30 inches. Hence, to find the distance of image from mirror, we have

$$\frac{1}{v} = \frac{1}{5} - \frac{1}{30} = \frac{1}{6},$$

or $\qquad v = 6.$

Therefore $v$ being positive, the image is real and inverted. It is 6 in. from the mirror, and consequently is diminished in proportion of 6 to 30, or 1 to 5.

*Ex.* 2. With the same mirror, to find the nature, position, and magnitude of the image, when the object is 4 inches from the mirror. In this case,

$$\frac{1}{v} = \frac{1}{5} - \frac{1}{4} = -\frac{1}{20},$$

or $\qquad v = -20.$

The image is therefore behind the mirror, and is virtual and erect. It is 20 inches from the mirror, and therefore is magnified in the proportion of 20 to 4, or 5 to 1.

*Ex.* 3. The focal length of a convex spherical mirror is 6 inches, find the nature, position, and magnitude of the image, when the object is 3 inches from the mirror.

In this case the focal length is negative, therefore,

$$\frac{1}{v} = -\frac{1}{6} - \frac{1}{3} = -\frac{1}{2},$$

or                      $v = -2.$

The image is therefore behind the mirror, and is virtual, and erect. It is 2 inches from the mirror, and therefore is diminished in proportion of 2 to 3.

## EXAMPLES.

1. The focal length of a concave mirror is 10 inches, what is the magnitude of the image of a small object 30 inches from the mirror?
Ans. one-half the object.

2. The focal length of a convex mirror is 8 inches, what is the position of the image of a small object 12 inches from the mirror?
Ans. 4⅘ inches behind the mirror.

3. Shew that the linear magnitudes of the image and object are as their respective distances from the *mirror.*

4. At what distance from a concave mirror must an object be placed that the image may be exactly one-half the object?   Ans. $3f.$

5. At what distance from a concave mirror must an object be placed that the image may be double the object?
Ans. $\frac{3f}{2}$.

6. At what distance from a convex mirror must an object be placed that the image may be half the object?   Ans. $f.$

7. At what distance from a concave mirror must an object be placed that the image may be *n* times the object?
Ans. $\frac{(n+1)f}{n}$.

8. At what distance from a convex mirror must an object be placed that the image may be 1-*n*th of the object?   Ans. $(n-1)f.$

9. The image of a candle placed 6 inches in front of a concave mirror is seen distinctly upon a screen 24 inches from the mirror; what is the focal length of the mirror?   Ans. 4⅘ in.

10. A luminous object is placed midway between a concave mirror and its focus; shew that the image will be double the object.

11. A pencil converging to a point midway between the mirror and its focus falls upon a convex mirror; shew that after reflection it will converge to a point whose distance from the mirror is equal to the focal length.

12. How far from a concave mirror must an object be placed that the image may be erect and three times the size of the object?   Ans. ⅔$f.$

13. A distinct image of a candle distant 6 feet from a screen is produced upon the screen by a concave mirror placed 4 feet from the candle; shew that a distinct image will also be produced upon a screen 8¾ feet from the candle if the mirror be placed 3 inches nearer to the candle.

### 154. Laws of refraction.—First. *The refracted ray lies in the same plane with the incident ray and the normal at the point of incidence.* Secondly. *If points be taken in the incident and refracted rays, equally distant from the point of incidence, and from these points perpendiculars be drawn to the normal at the point of incident, the ratio of these perpendiculars is constant for any given medium.*

This ratio is called the *refractive index*, and is commonly represented by the letter $m$.

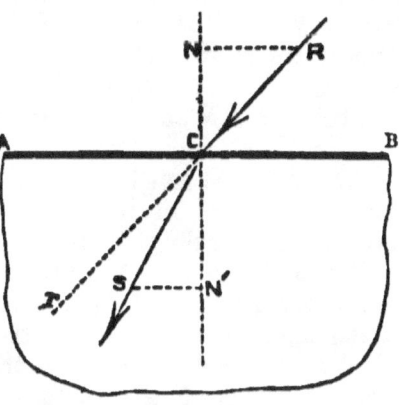

Let RC be a ray incident upon AB the surface of a refracting medium, and CS its path after refraction. Let R and S be equi-distant from C, and NN' the normal through C. Then, if RN and SN' be drawn perpendicular to NN', the ratio of RN to SN' is the same for all angles of incidence whatsoever, and if $m$ be the refractive index,

$$\frac{RN}{SN'} = m, \text{ or } RN = m \cdot SN'.$$

155. When light passes from a rarer into a denser medium, the value of $m$ is greater than unity, and consequently the angle of refraction is less than the angle of incidence. When light passes from a denser into a rarer medium, $m$ is less than unity, and the angle of refraction greater than the angle of incidence. Hence, by refraction, a ray of light is bent *towards* the perpendicular when passing into a denser medium, and *from* the perpendicular when passing into a rarer medium.

Between atmospheric air and plate or crown glass, $m = \frac{3}{2}$ nearly ; between atmospheric air and water $m = \frac{4}{3}$. When the light passes from glass or water into air, the refractive indices are the reciprocals of the values just given, and are therefore $\frac{2}{3}$ and $\frac{3}{4}$ respectively.

156. *To trace the path of the refracted ray when a ray of light is incident upon the surface of a given refracting medium.*

Let AB be the given surface, and *m* the refractive index.

Let RC be the incident ray. Through C draw the normal NN'. With C as a centre, and any distance CR as a radius, describe a circle. Draw RN perpendicular to NN', and take CD of such a length that RN = *m*. CD.* Through D draw DS parallel to CN', then CS is the path of the refracted ray. For draw SN' perpendicular to CN', then

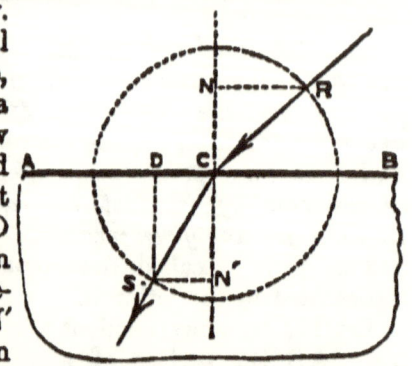

SN' = CD, and, consequently, RN = *m*. SN', which is the relation between RN and SN' required by the law of refraction.

157. **Critical angle. Total reflection.**—It has been seen that when a ray of light passes from a denser into a rarer medium, the angle of refraction is greater than the angle of incidence. Hence, as the greatest possible value of the angle of refraction is 90°, it follows that the greatest possible value of the angle of incidence will be less than 90°. This angle is called the *critical angle*. It is the angle of incidence when the angle of refraction is 90°, that is, when the refracted ray is parallel to the surface.

Thus, for example, let the ray RC, when passing out of a dense refracting medium, be refracted along CS, parallel to the surface AB, then the angle RCN is the critical angle.

Since RCN is the greatest angle at which refraction can take place for any ray within the medium, it follows that if a ray meet the surface at

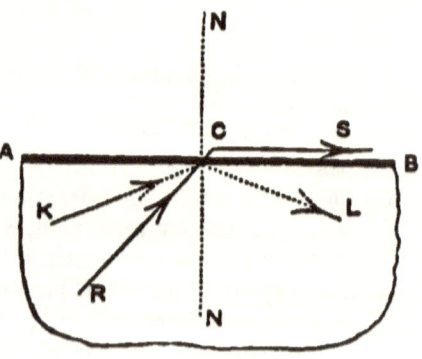

* If the medium be glass, CD must be taken so that RN = ⅔ CD, that is, CD must equal ⅔ of RN. Similarly, if the medium be water, CD must be taken equal to ¾ of RN.

an angle greater than the critical angle, a⁀, for instance, KC, it cannot emerge, but is reflected by the surface, in accordance with the law of reflection; and inasmuch as in this case no portion of the light is refracted, the reflection is said to be *total*.

The value of the critical angle for glass is about $41\frac{3}{4}°$; for water, about $48\frac{1}{2}°$.

### 158. Refraction of small direct pencils at plane surfaces.

—If a pencil of diverging rays fall upon a plane refracting surface, the refracted rays do not meet in any single point. If, however, the pencil be small, the rays of the refracted pencil diverge from a point so nearly that their distance from it is inappreciable. This point is the geometrical focus of the refracted pencil, and when the pencil is direct as well as small, its distance from the surface is $m$ times the distance of the focus of the incident pencil. When light passes from a rarer into a denser medium, $m$ is greater than unity, and consequently in this case the focus of the refracted pencil is more distant from the surface than the focus of the incident pencil. When light passes from a denser into a rarer medium, $m$ is less than unity, and consequently the focus of the refracted pencil is nearer the surface than the focus of the incident pencil.

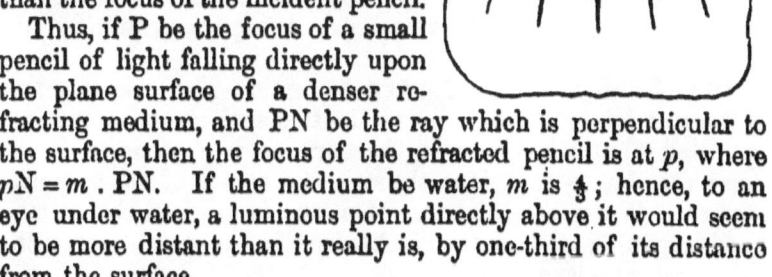

Thus, if P be the focus of a small pencil of light falling directly upon the plane surface of a denser refracting medium, and PN be the ray which is perpendicular to the surface, then the focus of the refracted pencil is at $p$, where $pN = m \cdot PN$. If the medium be water, $m$ is $\frac{4}{3}$; hence, to an eye under water, a luminous point directly above it would seem to be more distant than it really is, by one-third of its distance from the surface.

Again, let the light be passing from a denser into a rarer medium. Then, if P (fig. on next page) be the focus of the incident pencil, $p$ will be the focus of the refracted pencil, where, as before, $pN = m \cdot PN$, $m$, however, being now less than unity. If the medium be water, $m = \frac{3}{4}$. Hence, a luminous point under water would appear to an eye outside directly above it, to be less distant than it really is, by one-fourth of its distance from the surface.

**159. Lenses.**—Any portion of a refracting medium which has two opposite surfaces, either both spherical, or one spherical and the other plane, is termed a *lens*.

Lenses are of two classes, convex and concave. *Convex* lenses are those which are thickest at the centre. *Concave* lenses those which are thinnest at the centre.

Of convex lenses, those which have both surfaces convex are called *double convex* lenses (fig. 1); those which have one surface spherical and the other plane are called *plano-convex* lenses (fig. 2); and those which have one surface convex and the other concave (the radius of the CONCAVE surface being the greater) are called *meniscus lenses* (fig. 3). And, similarly, concave lenses are *double concave* when both surfaces are concave (fig. 4); *plano-concave* when one surface is spherical and the other plane (fig. 5); *convexo-concave* when one surface is convex and the other concave, the radius of the CONVEX surface being the greater(fig. 6).

**160. Axis of lens. Centre of lens.**—The *axis* of a lens is the line joining the centres of the two surfaces when both are spherical; but when one of the surfaces is a plane, it is the perpendicular to the plane through the centre of the spherical surface. When the thickness of a lens is so small that it may be disregarded, the *centre* of the lens is the point where the axis meets the lens.

As to thick lenses, if the lens is double convex, or double concave, with equal curvatures, the centre is that point in the axis which is midway between the two surfaces; if the lens be plano-convex, or plano-concave, the centre is the point where the axis meets the curved surface.

### 161. General effect of the two classes of lenses upon pencils of light.

—It is the property of convex lenses of every form to diminish divergency, of concave lenses to increase it. A diverging pencil is therefore rendered less divergent by a convex lens, and *may* become convergent; but by a concave lens it is rendered more diverging, and therefore can never become convergent. A parallel pencil is rendered convergent by a convex lens, but divergent by a concave lens. A converging pencil is rendered more convergent by a convex lens, but less convergent, and possibly parallel or divergent, by a concave lens. The general effect therefore of convex lenses is similar to that of concave mirrors, and of concave lenses to convex mirrors. It is important also to observe, that spherical lenses do not accurately refract large pencils of light. It is only when the incident pencil is small that the rays of the emerging pencil meet approximately in a focus. In the following sections therefore the pencils of light are always supposed to be small.*

### 162. Focus of lens. Focal length of lens.

—If a small pencil of *parallel* rays be incident *directly* upon a lens, the focus of the emergent pencil is called the principal focus of the lens (sometimes simply the focus of the lens), and the distance of this point from the lens is called the focal length. From what has been stated in the preceding article, it follows that the focus of every convex lens is real, and that the focus of every concave lens is virtual.

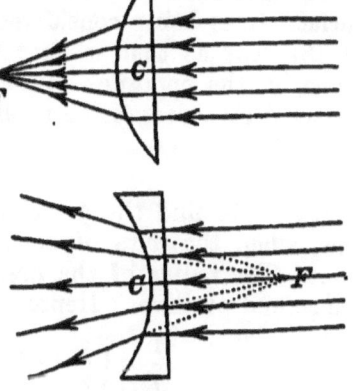

The focal length of any convex lens may be easily found by experiment. First cover the lens, leaving a small aperture in the central portion. Then allow the rays of the sun to fall upon the lens, and hold the lens in front of a screen, moving it towards or from the screen until

---

* It should be particularly observed, that in all the diagrams the pencils drawn are necessarily large. It would not be possible to draw small pencils with any approach to accuracy, and at the same time preserve distinctness in the diagram.

the position is found in which a distinct image of the sun is formed upon the screen: the distance of the lens from the screen is its focal length.

The focal length of a concave lens may be determined in the following way :—Take a convex lens, whose convexity is greater than the concavity of the given lens. The two together will then be equivalent to a convex lens, and the focal length of the combination may be experimentally determined as above. Then find the focal length of the single convex lens. *The difference of the reciprocals of these two focal lengths will be the reciprocal of the focal length of the concave lens.* For example, let the focal length of the convex lens be 2 inches, and that of the combination 5 inches. The difference of the reciprocals of these, or $\frac{1}{2} - \frac{1}{5}$ is $\frac{3}{10}$. The focal length of the concave lens is therefore $\frac{10}{3}$, or $3\frac{1}{3}$ inches.

163.* When the radii of the surfaces are known, and the refractive index, the focal length of a lens may be determined by the following formula :

$$\frac{1}{f} = (m - 1)\left(\frac{1}{r} - \frac{1}{s}\right)$$

Where $f$ is the focal length, $m$ the refractive index, $r$ the radius of the first surface of the lens, and $s$ the radius of the second surface, lines being considered positive when on the same side of the lens as that whence the light proceeds, and negative when on the opposite. When the value of $f$ is *negative*, the refracted rays pass *to* the focus, but when positive they proceed from it.

*Ex.* 1. To find the principal focus of a double convex glass lens, when the radii of both surfaces are equal. Here $m = \frac{3}{2}$ $s = r$. The radius of the first surface is negative, and that of the second positive. Hence,

$$\frac{1}{f} = \left(\frac{3}{2} - 1\right)\left(-\frac{1}{r} - \frac{1}{r}\right) = -\frac{1}{r}$$

$$\therefore \qquad f = -r.$$

That is, the focus is negative, and the focal length of the lens is equal to the radius of either surface.

* This may be omitted on a first reading.

*Ex.* 2. To find the principal focus of a plano-concave glass lens.

Here, as before, $m = \frac{3}{2}$. The radius of the first surface is positive. The second surface, being plane, may be regarded as a sphere of infinite radius. But if $s$ be infinite

$$\frac{1}{s} = 0. \quad \text{Therefore, in this case,}$$

$$\frac{1}{f} = \left(\frac{3}{2} - 1\right)\left(\frac{1}{r} - 0\right) = \frac{1}{2r}$$

$$\therefore \qquad f = 2r,$$

or the focal length is twice the radius of the concave surface.

**164. Focus of emergent pencil when a small pencil is incident upon a lens.**—When a small pencil of light is incident upon a lens, the position of the focus of the emergent pencil is determined by the following formula:

$$\frac{1}{v} = \frac{1}{f} + \frac{1}{u},$$

where $f$ stands for the focal length of the lens, $u$ for the distance of the focus of the incident pencil from the lens, and $v$ for the distance of the focus of the emergent pencil from the lens. If the focus of the incident pencil be in the axis of the lens, the focus of the emergent pencil will also be in the axis. When the focus of the incident pencil does not lie in the axis of the lens, the focus of the emergent pencil will lie in the line joining the focus of the incident pencil with the *centre* of the lens. This, however, is applicable only to the cases in which the distance of the focus of the incident pencil from the axis is small when compared with its distance from the lens. Such cases are alone considered in the present chapter. Let P (*fig.* Art. 165) be the focus of a small pencil incident upon the lens AB, whose centre is at C, then $p$, the focus of the emergent pencil, will lie in the line PC at a distance $pC$, as determined by the formula given above.

**165. Images formed by convex lenses.**—Let AB be the lens, whose focal length is 4 inches, and PQ the object, whose distance from the lens is 12 inches.

Since the lens is convex, the focal length is negative (Art. 162), therefore in this case $f = -4$, and for the pencil proceeding from any point in the object $u$ will be 12, therefore by the formula

$$\frac{1}{v} = -\frac{1}{4} + \frac{1}{12} = -\frac{1}{6},$$

$$\therefore \quad v = -6.$$

Hence, for the pencil proceeding from P, join PC, and in PC produced take $pC = 6$, then $p$ will be the focus of the refracted pencil. Similarly, $q$ will be the focus for the pencil proceeding from Q, and for all pencils proceeding from points between P and Q, the foci of the refracted pencils will lie between $p$ and $q$. Therefore $pq$ is the image of the object PQ. Since $pq$ and PQ are similar arcs of different circles they are as the radii. Therefore the image is to the object as $pc$ : PC ; that is, as 6 : 12, or as 1 : 2. The image, it will be seen, is inverted, and not being on the same side of the lens as the object, is *real*.

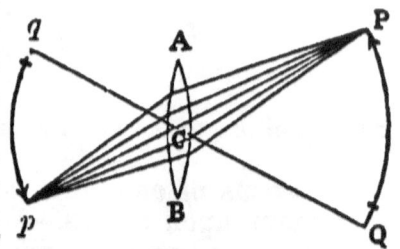

By proceeding similarly with objects placed at different distances from the lens, it will be found—

That when the object is at a greater distance than twice the focal length, the image is real, inverted, and diminished ; that when the object is at twice the focal length, the image is real, inverted, and of the same size as the object ; that when the object is distant less than twice the focal length, but more than the focal length, the image is real, inverted, and magnified ; and that when the object is distant less than the focal length, the image is virtual, erect, and magnified.

Hence also, as the object moves from a great distance up to the lens, the changes in the position of the image will be as follows :

When the object is at a great distance the image is at the focus ; as the object moves from infinity to a distance from the lens equal to twice the focal length, the image moves from the focus to twice the focal length ; as the object moves from twice to once the focal distance, the image moves from twice the focal distance to infinity ; and, lastly, as the object moves from the focal distance to the lens, the image (which is then virtual, and on the same side of the lens as the object) moves from infinity up to the lens.

166. **Images formed by concave lenses.**—Let $\dot{A}B$ be the lens, and its focal length 5 inches. Let PQ be the object, distant 8 inches from the lens. The lens being concave, the focal length is positive. (Art. 162.) Therefore, for determining the foci of the refracted pencil the formula gives,

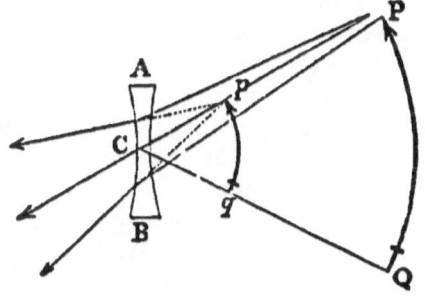

$$\frac{1}{v} = \frac{1}{5} + \frac{1}{8} = \frac{13}{40},$$

$$\therefore \qquad v = 3\tfrac{1}{13}.$$

Hence, for the pencil proceeding from P, join PC, and take $pC = 3\tfrac{1}{13}$, then $p$ is the focus of the refracted pencil. Similarly, $q$ will be the focus for the pencil proceeding from Q, and for all pencils proceeding from points between P and Q, the foci of the refracted pencils will lie between $p$ and $q$. Hence, the image $pq$ is virtual, erect, and diminished, the image being to the object as $3\tfrac{1}{13}$ to 8, or as 5 to 13.

A similar result will be obtained whatever the distance of the object from the lens. With concave lenses the image is always virtual, erect, and diminished.

167. **The use of lenses in correcting the defects of vision.**—The sensation of vision is produced by the light which enters the eye-ball through the small circular opening called the *pupil*. The size of this opening in the human eye varies somewhat, according to the intensity of the light, becoming greater or less as the light is less or more brilliant; but at its greatest expansion is only about ¼th inch in diameter. It hence results, that of the pencils of light which enter the eye, those only will have any sensible divergency which proceed from points comparatively near to the eye—say, within a distance of 3 feet. Light coming from all points beyond this distance will enter the eye in pencils which do not sensibly differ from parallel pencils. In the case, then, of the majority of objects, we see by means of parallel pencils; but in the case of those objects which lie within a range of about a yard, we see by means of diverging pencils. In its ordinary condition, the eye is suited to both these cases; it

can see by means of parallel pencils, and also by those which
have some sensible divergency, and possesses the power of passing
at will from the state in which it is suited to one kind of pencil
to that in which it is suited to the other kind. The most common
defect of vision is that which arises from the want of this power
of adaptation, so that the eye can see either by means of parallel
pencils only, or by means of diverging pencils only. The former
is the defect termed *long-sightedness*, and the latter, that termed
*near-sightedness.* The long-sighted person can see distinctly when
the object is sufficiently distant that the pencils which reach
his eye shall be sensibly parallel; but cannot see distinctly when
it is so near that the pencils have a sensible divergency. The
near-sighted person, on the other hand, can see distinctly when
the object is near, and sends diverging pencils to the eye; but
cannot see when it is so distant as to send parallel pencils.

In order, therefore, that a long-sighted person may see a near
object, the diverging pencils which proceed from it must be
changed into parallel pencils. This, as seen in Art. 161, can be
done by aid of *convex* lenses, and these, accordingly, are the
kind of glasses to be used.

In order that a near-sighted person may see a distant object,
the parallel pencils proceeding from it must be changed into
diverging pencils. This can be done (Art. 161) by aid of
*concave* lenses, and these, consequently, supply the remedy for
this defect of vision.

168. **The use of lenses in enlarging the range of
vision.**—Even with the most healthy eye the range of vision is
limited. An object may be too near or too distant to be distinctly
seen. In the former case the pencils are too divergent for the
eye to deal with; and in the latter case, the light is too weak to
excite the sensation of vision. Lenses are a convenient means of
enlarging the range of vision, in both directions; and when so
used, they form in the one case the MICROSCOPE, and in the
other the TELESCOPE.

The apparent magnitude of any object depends upon its
distance from the eye; the less the distance, the greater the
apparent magnitude, and *vice versa.* A small object may,
therefore, be magnified; that is, made to *appear* larger, by
simply bringing it nearer to the eye; and there would be no
limit to the degree in which we could thus magnify, if we could
see distinctly at all degrees of nearness. This, however, is not

the case. With all persons there is *a least distance of distinct vision.* Ordinarily this is about 8 inches; with long-sighted persons it is, as already seen, considerably more; with near-sighted persons it is less, being about 4 or 5 inches only. When an object is nearer than the least distance of distinct vision, the pencils of light which reach the eye are too divergent for use. A lens that will sufficiently diminish this divergency, will enable us to see distinctly. Convex lenses alone diminish divergency; these, therefore, are the kind of lenses to be used for this purpose, and when so used form what is termed a SIMPLE MICROSCOPE. In other words, a simple microscope is a convex lens, used for seeing an object when nearer to the eye than the least distance of distinct vision; and the object appears magnified, not because any magnified image has been formed by the lens, but simply because it is seen at a less distance from the eye than ordinarily. The magnifying power of a simple microscope depends upon the degree of nearness at which it enables us to see distinctly. It also depends upon the least distance of distinct vision of the person who uses it. Thus, if the least distance of distinct vision be 8 inches, and the microscope enables us to see at $\frac{1}{2}$ inch, the magnifying power is $8 \div \frac{1}{2}$, or 16.

When a higher magnifying power is needed than that which can be conveniently obtained by a simple microscope, a combi-nation of lenses, forming a COMPOUND MICROSCOPE, is used. The essential parts of this instrument are two convex lenses, one of which, called the *object-glass,* is used for forming a magnified image of the object, and the other, called the *eye-glass,* is used for viewing this image at a distance from the eye less than the least distance of distinct vision, and so still further increasing its apparent magnitude.

Thus, let O be the object-glass, and E the eye-glass. Let P be any small object placed a little farther from O than its focal length, then (Art. 165) a magnified im-age will be formed on the opposite side of the lens. Let Q be this image. Then the eye-glass E being placed at the proper distance for distinct vision, the object will be seen by an eye behind it, and will appear highly magnified.

The TELESCOPE, like the compound microscope, consists of an

object-glass and an eye-glass.  Optically considered, the two
instruments differ only in the form and size of the object-glass.
In the microscope the object glass is small, and of small focal
length.   In the telescope the object-glass is large, and of large
focal length.  This difference arises from the different uses of
the object-glass in the two instruments.   In the telescope, the
use of the object-glass is not, as in the microscope, to form a
magnified image of the object, but to form a *bright* image ; and
in order to this, it is necessary that a large quantity of light
should fall upon the object-glass from each point in the object :
and it is clear, that the larger the object-glass, the larger is the
quantity of light which falls upon it.   Moreover, the object-
glass is not only large, but of large focal length.  This is
necessary in order that the pencils of light falling upon it may
be the more accurately refracted, and so produce a brighter
image.   The image, as in the microscope, is viewed by aid of
the eye-glass.

## EXAMPLES.

1. A convex lens of .6 inches focal length is combined with a concave
lens ; the focal length of the combination is 24 inches, what is the focal
length of the concave lens ?                          Ans. 8 inches.
2. A small pencil of light, proceeding from a point distant 30 inches
from a convex lens, is refracted to a point distant 6 inches, determine the
focal length of the lens.                             Ans. 5 inches.
3. Find the focal length of a glass meniscus, the radii of whose surfaces
are 4 and 5 inches.                                   Ans. 40 inches.
4. Find the focal length of a double concave glass lens, the radius of
each surface being 3 inches.                          Ans. 3 inches.
5. Find the focal length of a double convex lens, formed of a substance
whose refractive index is $\frac{3}{2}$, the radius of each surface being 5 inches.
                                                      Ans. 4½ inches.
6. If the radius of the spherical surface of a plano-concave lens be 6
inches, what must be the refractive index of the material of the lens, in
order that its focal length may be 10 inches ?        Ans. $\frac{3}{5}$.
7. If an object be placed 10 inches from a convex lens, whose focal
length is 6 inches, find the position of the image.
                                         Ans. 15 inches behind the lens.
8. If an object be placed 10 inches from a convex lens, whose focal
length is 12 inches, find the position of the image.
                                         Ans. 5 feet in front of the lens.
9. How far in front of the convex lens, whose focal length is 1 inch,
must an object be placed in order that the image may fall 6 inches behind
the lens ?                                            Ans. 1⅙ in.
10. How far from the concave lens, whose focal length is 3 inches, must
an object be placed in order that the image may be 2 inches from the
lens ?                                                Ans. 6 inches.
11. A near-sighted person can read most distinctly when a book is at

6 inches from his eyes, what must be the focal length of the glasses that will enable him to read distinctly at 20 inches ?          Ans. 8⅓ in.

12. If the person referred to in the preceding wish to see distinctly at a greater distance than 20 inches, will he require glasses of a longer or of a shorter focal length than that just determined ?

13. If a small object and its image be at distances of 6 and 2·5 inches on opposite sides of the lens, find the focal length of the lens. *U. of L. Matriculation, 1869.*

14. If a candle is placed at a distance of 6 feet from a wall, and a distinct image of the flame is produced upon the wall by a lens held at 1 foot from the candle, show that a distinct image will also be produced when the lens is at 5 feet from the candle ; and compare the sizes of the two images. *U. of L. Matriculation, 1872.*

15. If a real image 5 times as high as the object is to be thrown on a screen at a distance of 36 inches from the object, show what must be the focal length of the lens employed. *U. of L. Matriculation, 1873.*
                                                                        Ans. 5 in.

16. The sun is 90,000,000 miles distant, and his diameter is (say) 900,000 miles. What will be the apparent diameter of the sun's image as given by a spherical mirror of 10 feet radius ? *U. of L. Examination for Women, 1875*          Ans. 0·6 in.

17. The focal length of a plano-convex lens is 10 inches. A small candle-flame is placed on the axis of the lens at a distance of 25 inches from it. Describe by means of a diagram the nature and position of the image of the flame formed by the lens. *U. of L. Examination for Women, 1876.*

18. A ray of light passes from air into glass, the refractive index of glass with regard to air being 1·5. Given the angle of incidence at the common surface, draw a diagram to show how the angle of refraction may be accurately determined. *U. of L. Matriculation, 1876.*

19. A small object 0·1 inch long is placed at the distance of 3 feet from a convex glass lens, of 12 inches focal length. What is the length of its image and the distance of this from the lens ? *U. of L. First B. Sc., 1876.*
                                                        Ans. 0·05 inch. and 18 inches.

# CHAPTER IX.

169. **Heat** is the name given to the unknown cause of the sensation of warmth. The ordinary means by which this sensation is produced are: (1) The action of the sun, (2) the combustion of fuel, (3) the digestion of food, (4) physical exercise, (5) friction, or the rubbing together of solid substances. These are matters of universal observation. We warm ourselves by the rays of the sun, by burning coals or wood, by the meals we eat, and by vigorous exercise. We rub our hands when cold, and by the friction of the axles of our railway carriages the wheels are sometimes set on fire. Heat is produced also by electricity, as is seen when trees or buildings are struck by lightning; and by chemical action, as when sulphuric acid is mixed with water.

The ordinary SOURCES OF HEAT are therefore (1) the sun, (2) fuel, (3) food (which indeed, so far as concerns heat, is fuel burnt in the bodies of animals), and (4) mechanical effort. There is however reason for thinking that heat is but a form of force; other forms of which are gravitation, light, electricity, and magnetism. And if this be so, then it may be said that heat is everywhere, and that all existing force under whatever form it may appear is a possible source of heat.

170. Though primarily we think of heat as the cause of a certain sensation, our sensations are not sufficiently precise to be in themselves a measure of its degree or quality. Different persons are differently affected by the same substance; to one it seems warm, to another cold. Even in the case of the same person, the same substance may excite a feeling of warmth in his right hand but of cold in his left. We have therefore to seek for some more accurate measure of heat, and this we find in one of its most obvious effects; namely, the expansion which

is caused in the volume of a body by an increase of heat. That in ordinary circumstances heat expands bodies is easily seen by simple experiments. A metal rod, for instance, that just fits into a ring when cold will, if heated in the fire, no longer pass through the ring; and if its length be similarly tested it will be found longer than before. If a glass tube containing water or spirit be held in the flame of a lamp, the fluid will quickly rise in the tube. If, instead of using a lamp, we plunge the tube into different vessels containing heated water, the liquid will in like manner rise in the tube, and will rise the higher the hotter the water is. In like manner diminution of heat will ordinarily lessen the volume of a body. If the tube be placed in a liquid colder than itself, the liquid in the tube will fall; and if placed successively in different liquids, will fall the lower in that which is the colder.

171. **Temperature.**—*Two bodies are said to be of the same temperature, when the change of volume they produce in any small\* body placed in contact with them is in all respects the same.* Thus, if a small quantity of mercury or of spirits of wine be enclosed in a narrow glass tube, only partially filling it, and if when this tube is placed separately in connection with two different bodies, the liquid in each case rise from the same point A to the same point B, the two bodies are said to have the same temperature; but if in one case the liquid rise from A to B, and in the other from A to C, a point above B, the second of the two bodies is said to be of higher temperature than the first.

The temperature of a body is independent of its mass. If from a vessel containing any quantity of liquid a part be taken, say one-half or one-third, and this be separately tested by such a tube as that already used, the temperature of the part will be found to be the same as that of the whole; that is to say, the change of volume of the liquid in the tube will be the same when placed in the half or the third as when placed in the whole. In the application of this test, care must of course be taken that in separating the part no loss of heat arise from contact with any cooler body or from any other cause.

But though the temperature of the whole and of the part be

---

\* By "small" is here meant a body whose weight is inconsiderable in comparison with the weight of the bodies to be examined. The reason for this proviso will be seen further on.

the same, it is clear that the quantity of heat in the whole is greater than the quantity of heat in the part, and this in the exact proportion of the whole to the part ; that is, there is twice as much heat in the whole as in the half, three times as much as in the third, and so on. Hence, while the temperature of a body is independent of its mass, the quantity of heat varies directly as the mass when the temperature is constant. At present the comparison is instituted only between the quantity of heat in any given mass and that in another mass of the *same* body. For instance, if two pieces of iron, the one weighing 2 lbs. and the other 3 lbs., be of the same temperature, then the quantity of heat to the former is to that in the latter as 2 to 3. If different substances be taken the same cannot be affirmed ; and, as will be seen hereafter, although the weights be the same and the temperatures also the same, the quantity of heat in the one may differ considerably from the quantity of heat in the other.

172. **Thermometers** are instruments for the accurate measurement of temperature. The most common of these is the mercurial thermometer. It is thus constructed : A glass tube with a fine uniform bore * has a globular or elongated bulb blown at one end. The bulb is heated, and by the consequent expansion of the air contained in it some part of the air is driven out. The open end is then placed in a vessel containing mercury, and as the instrument cools the expansive force of the enclosed air diminishes, and becoming less than the atmospheric pressure upon the mercury in the open vessel, some portion of mercury is driven up the tube, and if the bulb were sufficiently heated some of the mercury will be driven into the bulb. If necessary the process must be repeated until the mercury fill the bulb and part of the tube. Heat is then applied to the tube, so that the upper part of the mercury may boil, and its vapour may exclude all the air and moisture in the tube. When this is effected the open end of the tube is hermetically sealed up by means of a blow-pipe.

* The uniformity of the bore may be tested thus : A small quantity of mercury is introduced into the bore such as fills a certain length of the bore, say half an inch. If the bore be uniform the length of the column of mercury will still be half an inch, in whatever part of the tube the mercury may be placed. By passing it therefore along the tube, and observing whether there be any variation in the length of the column, the uniformity of the bore may be very accurately tested.

**173. Graduation of the thermometer.**—The instrument prepared as above is first placed in a vessel containing pounded ice in a melting state, with an opening at the bottom by which the water can drain off. The point at which the mercury stands in the tube is then marked, and is commonly called the freezing-point.

The instrument is next placed in the steam of water boiling under the mean pressure of the atmosphere. The point at which the mercury then stands is also marked, and is commonly called the boiling-point.

In the *Centigrade Thermometer* the freezing-point is marked o, and the boiling-point 100, and the interval is divided into 100 parts or degrees.

In the *Reaumur Thermometer* the freezing-point is marked o, and the boiling point 80; and the interval is divided into 80 parts.

In the *Fahrenheit Thermometer* the freezing-point is marked 32, and the boiling-point 212; and the interval is divided into 180 parts.

In each the degrees thus obtained are continued below the freezing-point, and temperatures below zero on either scale are denoted by the minus sign. Thus − 10° C means 10 degrees centigrade below the freezing-point; − 10° F means 12 degrees Fahrenheit below the zero of this scale, or 44 degrees below the freezing-point.

174. Since 100 degrees centigrade are equal to 180 degrees Fahrenheit, it will be seen that a degree centigrade is to a degree Fahrenheit in the ratio of 9 to 5. Hence, to reduce any temperature from the centigrade to the Fahrenheit scale : First multiply the given temperature by 9 and divide by 5, this gives the required number of degrees F reckoned from the freezing-point; then add 32, and the required number on the F scale is obtained.

Conversely to reduce from the Fahrenheit to the centigrade scale : First subtract 32, this gives the number of degrees F above the freezing-point; then multiply by 5 and divide by 9, and the required number on the C scale is obtained.

These results may be expressed by the equations

$$F = \frac{9}{5} C + 32 \qquad\qquad C = \frac{5}{9} (F - 32.)$$

Strictly speaking, 100° C and 212° F are not identical. In instruments constructed with great accuracy for scientific purposes, the boiling-point of the C thermometer marks the temperature of boiling water under a pressure of 76 centimetres of mercury at the freezing-point in the latitude of Paris. The boiling-point of the F thermometer marks the temperature of boiling water under the pressure of 29·905 inches of mercury at the freezing-point in the latitude of London. The temperature denoted by 100° C is slightly in excess of that denoted by 212° F, but the difference is so small that it may ordinarily be neglected.

### EXAMPLES.

1. Reduce   25° C to the F scale.                    Ans.   77° F
2. Reduce   86° F to the C scale.                    Ans.   30° C
3. Reduce – 13° F to the C scale.                    Ans. – 25° C
4. Reduce – 20° C to the F scale.                    Ans. –  4° F
5. Reduce   30° C to the R scale.                    Ans.   24° R
6. Reduce   28° R to the C scale.                    Ans.   35° C
7. Reduce   28° R to the F scale.                    Ans.   95° F
8. Reduce  140° F to the R scale.                    Ans.   48° R
9. What temperature is expressed by the same number in both the C and F scales?                                      Ans. – 40°.

175. In the graduation of the mercurial thermometer, by dividing the interval between the freezing and boiling-points into *equal* parts, and by giving to these parts the common name of a degree, it is assumed that equal increments of temperature give rise to equal increments in the volume of the mercury. This, though not absolutely true, is found to be practically so within the limits mentioned, and to a certain range beyond them. The beginner may perhaps be led to enquire how this can be known, inasmuch as ordinary methods for testing the expansion of bodies at different temperatures presuppose the possession of a thermometer, and to attempt to shew that the expansion of mercury is uniform by a thermometer constructed as above would be to reason in a vicious circle. It is possible however in some cases to determine the temperature without the use of a thermometer, and thence to test the accuracy or other-

wise of the assumption upon which the thermometer has been constructed. Thus, for instance, if equal quantities by weight of water at o° C and at 100° C be mixed without loss of heat, the resulting temperature of the mixture will be 50° C. If 3 parts by weight of water at o° C be mixed with 1 part of water at 100° C, the temperature of the mixture will be 25° C. By taking the proper proportions any required temperature between the freezing and boiling-points may be obtained, and this can be examined by the thermometer. If in each case the temperature denoted by the thermometer agrees with that obtained by calculation, the accuracy of the assumption upon which the thermometer was constructed is so far demonstrated.

**176. Limits to the use of the mercurial thermometer.** —Mercury solidifies at about $-39°$ C ($-38°$ F), and when approaching this temperature its expansion is irregular; it is therefore not available for temperatures lower than $-36°$ C. Again, mercury boils at about 350° C (662° F), and hence is not available for temperatures higher than this.

**177. The spirit thermometer.**—Alcohol has not solidified at any degree of cold hitherto attained, and hence a thermometer containing alcohol instead of mercury is a convenient instrument for the measurement of low temperatures. It is graduated by comparison with a mercurial thermometer, being placed in baths of different temperatures, and the temperature recorded in each case by the mercurial thermometer being then marked upon the spirit thermometer. In the higher part of the scale thus obtained equal increments of temperature are not marked by equal increments of the spirit, but in the range of the lower temperatures measureable by the mercurial thermometer the expansion of the spirit is approximately uniform. The length for one degree thus obtained is continued downward for temperatures lower than the freezing-point of mercury.

**178. Correction of the mercurial thermometer.** — However carefully a mercurial thermometer may be constructed it is found to be liable to several sources of error.

The most important of these is the change in the zero point which is found to take place after the lapse of time. The instrument when placed in melting ice no longer stands at o° C, but at some point above it, perhaps at 1° or 2°. All the readings given by such a thermometer are consequently too high. The extent of the error arising from this source may be diminished by keeping the tube after it has been filled a long time before graduating it, and by well annealing the glass. In an instrument so constructed the error will not perhaps amount to more than three or four tenths of a degree centigrade. Such an error may be disregarded in ordinary cases, but where great accuracy is required the thermometer must be first tested to determine its precise amount of error.

A second source of error is the temporary displacement of the zero point when the thermometer has been used for high temperatures. In cooling the bulb and stem do not for a while contract to their original volumes, and hence the readings given are lower than the correct reading. The amount of error arising from this source is ordinarily not more than one-tenth of a degree, and it disappears after the lapse of about ten or twelve days.

A third source of error arises from the unequal expansion of the bulb and stem when the bulb is immersed in a medium of high temperature and the stem exposed to the air at a much lower temperature.

The correctness of the reading given by a thermometer may also be affected by the position in which it is held. A thermometer which has been graduated for the upright position will give slightly higher readings when used in a horizontal position; and conversely, a thermometer graduated for the horizontal position will give slightly lower readings when used in a vertical position, the density of the mercury being slightly greater when the column is vertical than when it is horizontal.

179. **Thermal unit.**—Having learnt how to measure *temperature*, it is needful to adopt some standard by which to measure *quantity of heat*. The standard commonly taken is the quantity of heat necessary to raise a weight-unit of water from $0°$ C to $1°$ C.

The unit of weight is sometimes a pound avoirdupois, sometimes a kilogramme, sometimes a gramme. In many questions it is a matter of indifference which of these be taken, provided only the same unit be used throughout.

The quantity of heat necessary to raise a weight-unit of water from $0°$ to $t°$ is very nearly, though not exactly, $t$ times as much as that required to raise it from $0°$ to $1°$, and hence the thermal unit may be defined as *the quantity of heat necessary to raise a weight-unit of water one degree in temperature.*

180. **The coefficient of expansion** is *the ratio which the increment of volume, given to any body by an increase of $1°$ in temperature, bears to the volume of the body at freezing-point.* It may also be defined as *the increment given by an increase of $1°$ in temperature to any body whose volume at the freezing-point is unity.*

In most cases this ratio is a variable quantity ; that is to say, the increment of volume arising from an increase of $1°$ in temperature is different at different temperatures, the increment, for instance, arising from a change from $60°$ to $61°$ being different from that arising from a change from $50°$ to $51°$.

In some cases, however, the coefficient of expansion is a constant quantity, either absolutely or approximately, and every increase of $1°$ in temperature adds the same amount to the volume. Thus, if the coefficient of expansion be $\frac{1}{500}$, then 500 units at freezing-point will become 501 at $1°$, 502 at $2°$, 510 at $10°$, and so on. Hence when the coefficient of expansion is constant, the relation between the volume at freezing-point and the volume at any given temperature may be readily found.

Let $a$ be the coefficient of expansion, and $V_0$ the volume at freezing-point, then

$$\text{increment for } 1° = aV_0$$
$$\text{do.} \quad \text{for } 2° = 2aV_0$$
$$\text{do.} \quad \text{for } t° = atV_0 ;$$

and hence, if $V$ be the volume at $t°$,

$$V = V_0 + at\,V_0 = V_0\,(1 + at),$$

or the volume at $t°$ is the volume at freezing-point *multiplied* by $1 + at$ ; and conversely the volume at freezing-point is the volume at $t°$ divided by $1 + at$.

Hence also if $V_1$ and $V_2$ be the volumes which any substance assumes at the temperatures $t_1$ and $t_2$, we have this simple relation :

$$\frac{V_1}{1 + at_1} = \frac{V_2}{1 + at_2},$$

since each of these equal quantities represents the volume at freezing-point.

. L

It must be specially noted that in this formula, $t_1$ and $t_2$ denote the number of degrees above the freezing-point, whatever the scale that may be employed; and hence, if the F scale be used, 32 must be subtracted from the number by which the temperature is described.

*Ex.* 1. If 120 cubic units at 20° C be raised to 40° C, what will be the volume, the coefficient of expansion being ·004?
Let V be the volume required; then,

$$\frac{V}{1 + ·004 \times 40} = \frac{120}{1 + ·004 \times 2}$$

$$V = \frac{120 \times 1·16}{1·08} = 128\tfrac{8}{9}.$$

*Ex.* 2. If 160 cubic units at 40° F be raised to 60° F, what will be the volume, the coefficient of expansion being ·003?
Let V be the volume required; then,

$$\frac{V}{1 + ·003 \times 28} = \frac{160}{1 + ·003 \times 8};$$

$$\therefore \quad V = \frac{160 \times 1·084}{1·024} = 106\tfrac{7}{8}.$$

### EXAMPLES.

1. If 100 cubic units at 20° C be raised to 30° C, what will be the volume, the coefficient of expansion being ·003?    Ans. 102·83

2. If 212 cubic units at 50° F be raised to 104° F, what will be the volume, the coefficient of expansion being $\frac{1}{180}$?    Ans. 248

3. To what temperature must 120 cubic units at 20° C be raised that the volume may become 125, the coefficient of expansion being $\frac{1}{111}$?    Ans. 41° C.

4. If 100 cubic units at 10° C become 120 when raised to 60° C, what is the coefficient of expansion?    Ans. $\frac{1}{250}$.

5. If 50 cubic units at 40° C be raised to 80° C, what is the increase of volume, the coefficient of expansion being $\frac{1}{160}$?    Ans. 5.

6. If 240 cubic units at 100° F be lowered to 60° F, what is the diminution of volume, the coefficient of expansion being $\frac{1}{112}$?   Ans. 24.

181. It is sometimes convenient to know the linear expansion of bodies, and their superficial expansion, as well as their expansion in volume. As the coefficient of expansion is in all cases a small fraction, these three quantities have a very simple relation to each other. Thus if $x$ be the coefficient of linear

expansion, then a unit of length becomes $1 + x$ when raised from the freezing-point to one degree above it. Consequently a unit of surface becomes $(1 + x)^2$, or $1 + 2x + x^2$; that is, $1 + 2x$ nearly. Hence the coefficient of superficial expansion is $2x$, or twice the coefficient of linear expansion. Similarly a unit of volume becomes $(1 + x)^3$, or $1 + 3x + 3x^2 + x^3$; that is, $1 + 3x$ nearly. Hence the coefficient of cubic expansion is $3x$, or three times the coefficient of linear expansion.

N.B. The expansion in the content of any vessel is the same as the expansion of a corresponding volume of the substance of which the vessel is composed.

182. **Expansion of solids.**—Solids are not uniform in their expansion, but are approximately so between $0°$ C and $100°$ C. As the temperature increases, the rate of expansion increases, and the more rapidly the nearer the temperature approaches the melting-point. The rate of expansion, moreover, is different for different substances. The following table shows the *linear* expansion of the several substances named, between $0°$ and $100°$ C, as determined by some experimentalists.

| | |
|---|---|
| Glass............... | ·0000084 |
| Platinum ......... | ·0000088 |
| Iron ............... | ·0000118 |
| Brass............... | ·0000188 |
| Silver ........... | ·0000191 |
| Tin ............... | ·0000217 |
| Lead ............... | ·0000288 |

N.B. The coefficients here given are for $1°$ C.; if the coefficients of expansion for $1°$ F. be required, they can be found by taking five-ninths of these numbers.

183. **Compensation pendulum.**—The time of the vibration of a pendulum is dependent upon its length, and hence the uniformity of a clock is disturbed by the changes in the length of the pendulum arising from changes in the temperature. To remove this source of error, advantage has been taken of the fact that different metals expand at different rates, and the compensation pendulum is an arrangement of rods of steel and brass so placed that the expansion of one metal shall be neutralized by that of the other. Thus let AB and EF be rods of steel, and DC a rod of brass. Suppose that in conse-

quence of a change of temperature AB expand to A*b*, CD to
*cd*, and EF to *e*F, then if

$$Cc + Dd = Bb + Ee,$$

or the expansion of the brass
rod be equal to that of the steel
rods, the position of F, the centre
of the bob, will remain un-
changed. Now it will be seen
from the table given above that
the expansion of brass is to that
of steel as 94 to 59, and consequently such an arrangement
as is here represented is impossible, since it would require the
expansion of the brass to be
rather more than double that
of the steel. That the expansions
of the two metals may be equal,
the length of the brass rod must
be to that of the iron as 59 to
94, and the accompanying figure
exhibits an arrangement by
which such a proportion may
be conveniently secured, where
the effective part of the instru-

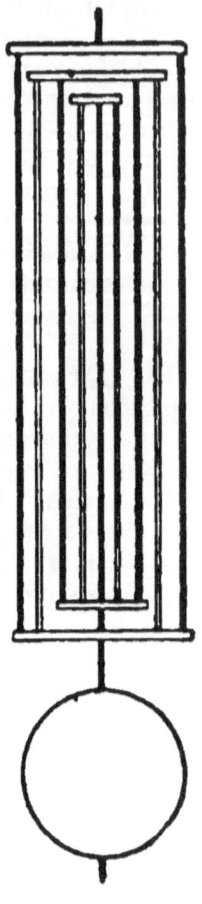

ment consists of three steel rods (represented
by the dark lines) and two brass rods (represented by the open lines).

184. **Expansion of liquids.** *Real and apparent expansion.*—When a liquid is heated expansion takes place, not only
in the liquid itself, but also in the vessel containing it. If the
latter be disregarded, the expansion noted is the apparent expansion of the liquid, and is less than the real expansion by the
expansion of the content of the vessel. If, for instance, a liquid
be heated in a graduated tube, such as a thermometer tube, and
the increase be measured by noting the number of degrees over
which it rises; or if a vessel of known content be filled, and
the increase be measured by taking the weight of the portion
that overflows when the vessel is heated; in either case we have

only the apparent expansion; for the vessel itself having ex panded during the process, a part of the expansion of the liquid is employed in filling up the increased volume of the vessel. Hence,

real expansion = apparent expansion + expansion of vessel.

If then the coefficient of expansion of any convenient solid such as glass can be found independently, the real of expansions of all liquids that have no action upon glass can be readily found; and conversely, whenever the real expansion of any liquid such as mercury is known, we can, by observing the apparent expansion of that liquid when heated in any vessel upon which it exerts no action, determine the expansion of the substance of which that vessel is made; for

exp. of vessel = real exp. of liquid — apparent exp. of liquid.

It is by this method that the coefficient of expansion of many solids has been determined by some experimentalists.

185. One method by which the expansion of liquids may be found, independently of any knowledge of the expansion of the containing vessel, is by means of a U tube. One limb of the tube is surrounded by a vessel containing ice, and the other by a vessel containing water of some known temperature, $t°$ C. The U tube is partly filled with the liquid to be examined, and when the liquid in the one limb has fallen to the temperature $o°$ C, and the liquid in the other limb has risen or fallen to the temperature $t°$ C, the height at which the liquid stands in each limb is carefully measured. Let these be $h$ and $h'$ respectively. Hence a column of $h$ units at $o°$ equals in weight a column of $h'$ units at $t°$, and therefore

$$h' - h = \text{linear expansion of } h \text{ units for } t°;$$

$$\therefore \quad \frac{h' - h}{h} = \text{linear expansion of } \text{i unit for } t°;$$

$$\therefore \quad \frac{h' - h}{ht} = \text{coefficient of linear expansion.}$$

186. The expansion of mercury as determined by the method just described, as also by other methods, is found to be so nearly uniform throughout a considerable range of temperature, that, except where extreme accuracy is desired, it may be treated as

constant, the mean coefficient of expansion being the same as far as the fifth place of decimals, namely ·00018, through so wide a range as from 0° C to 300° C.

187. Having found the coefficient of expansion of mercury, that of glass may be found, as indicated in Art. 184, by taking the difference between the real expansion of mercury and its apparent expansion in a glass vessel.

188. The expansion of a liquid may also be found by weighing a solid body, say a piece of glass, in the liquid at 0°, and at a known temperature $t°$, and observing the loss of weight in each case.

Let $l_1$ and $l_2$ be the observed losses in weight, and V the volume of the solid at 0°, then

$$l_1 = \text{weight of V units of the liquid at } 0°;$$

and if $a$ be the coefficient of expansion of the solid,

$$l_2 = \text{weight of V}(1 + at) \text{ units of the liquid at } t°.$$

Let $x$ be the volume which V units of the liquid at 0° would attain by being raised in temperature to $t°$, then since at the same temperature volumes are as the weights,

$$x : V(1 + at) :: l_1 : l_2,$$

or

$$x = \frac{V l_1 (1 + at)}{l_2};$$

therefore the increment of V units for $t°$ is

$$\frac{V l_1 (1 + at)}{l_2} - V,$$

and the increment of 1 unit for $t°$ is

$$\frac{l_1 (1 + at) - l_2}{l_2};$$

$$\therefore \quad \text{coeff. of exp.} = \frac{l_1 (1 + at) - l_2}{l_2 t}.$$

189. The expansion of liquids is, speaking generally, variable, and for the most part increases with an increase of temperature. Water is found to be at its greatest density at 4° C, and from this point it expands both by increase and by diminution of temperature.

As the temperature rises from 4° C, the rate of expansion increases with the increase of temperature. Thus the mean coefficient of expansion

from   4° C to   10° C is ·000047
"   10° C "   20° C " ·000152
"   20° C "   30° C " ·000254
"   50° C "   60° C " ·000493
"   90° C "   100° C " ·000749

As the temperature falls below 4° C, the rate of expansion increases with the diminution of temperature. Thus the mean coefficient of expansion

from   4° C to   0° C is ·000032
"   0° C "   −5° C " ·000114
"   −5° C "   −9° C " ·000233

**190. Expansion of gases.**—Except when near to the point at which they pass into the liquid state, gases are found to expand uniformly under an increase of temperature, and the rate of expansion is practically the same for all gases.

On the C scale, the coefficient of expansion is ·003665, or $\frac{1}{273}$.

On the F scale, the coefficient of expansion is ·002036, or $\frac{1}{491}$.

..ence the change in the volume of a gas, so far as it arises from a change of temperature, may be readily found, as shown in Art. 180.

*Ex.* 1. What will be the volume of 10 litres of gas at 40° C when raised to 60° C?

Let V be the required volume, then

$$\frac{V}{1 + \frac{60}{273}} = \frac{10}{1 + \frac{40}{273}} ;$$

$$\therefore \qquad V = \frac{10 \times 333}{313} = 10·64 \text{ litres, nearly.}$$

*Ex.* 2. To what temperature must 100 cubic feet of gas at 40° F be raised that its volume may be 120 cubic feet?

Let *t* be the required temperature, then

$$\frac{120}{1 + \frac{t - 32}{491}} = \frac{100}{1 + \frac{40 - 32}{491}} ;$$

$$\therefore \quad \frac{120}{459+t} = \frac{100}{499};$$

$$\therefore \quad 459+t = \frac{499 \times 120}{100}; \quad \therefore \ t = 139°\cdot 8 \ \text{F.}$$

191. As seen in Art. 130, the volume of a gas changes if the pressure upon it changes, and by the preceding Article it changes also if the temperature changes. When both temperature and pressure change, the resulting change of volume may be found by first finding the volumes which in each case the g s will have at the freezing-point; and then, the temperature being the same, the volumes will, by Boyle and Mariotte's law, be inversely as the pressures.

Thus if a certain quantity of gas have the volume $V_1$ under a pressure $p_1$ at a temperature $t_1°$ C, and the volume $V_2$ under a pressure $p_2$ at a. temperature $t_2°$ C, then if $a = \frac{1}{273}$, the volumes at $0°$ C are

$$\frac{V_1}{1+at_1} \ \text{and} \ \frac{V_2}{1+at_2}$$

respectively, and

$$p_1 : p_2 : : \frac{V_2}{1+at_2} : \frac{V_1}{1+at_1}.$$

Ex. If 13 litres of air have a temperature $7°$ C when the barometer is at 74 centimetres, what will be the volume at the temperature $42°$ C, the barometer standing at 78 centimetres?

Let V be the volume, then reducing both to $0°$ the volumes will be

$$\frac{13}{1+\frac{7}{273}} \ \text{and} \ \frac{V}{1+\frac{42}{273}}$$

respectively, and therefore

$$74 : 78 : : \frac{V}{1+\frac{42}{273}} : \frac{13}{1+\frac{7}{273}}$$

$$: : \frac{V}{315} : \frac{13}{280};$$

$$\therefore \quad \frac{78\,V}{315} = \frac{13 \times 74}{280};$$

$$\therefore \quad V = \frac{13 \times 74 \times 315}{78 \times 280} = 13\tfrac{7}{8} \ \text{litres.}$$

## EXAMPLES.

1. If 30 litres of gas are cooled from 25° C to 0° C, what is the diminution in volume, the pressure being constant ?  Ans. 2·51 litres.

2. If 20 cubic feet of air at 30° C be raised to 90° C, what is the increase in volume, the pressure being constant ?  Ans. 3·96 cub. ft.

3. If a litre of air at 0° C weigh 1·293 grammes when the barometer is at 760 mm., what will be the weight of a litre of air at 27° C, the barometer standing at the same height ?  Ans. 1·177 grm.

4. What will be the weight of a litre of air at 42° C when the barometer stands at 735 mm. ?  Ans. 1·084 grm.

5. If a cubic foot of air weigh 1¼ oz. when the temperature is 60° F, and the barometer is at 30 inches, what will be the weight of the air in a room 26 ft. long, 15 ft. wide, and 8 ft. high, when the temperature is at 35° F, and the barometer at 28·5 inches ?  Ans. 243 lbs. 4½ oz.

6. Air at a temperature 15° C is enclosed in a vessel and heated to 93° C, compare the elastic force of the enclosed air with that of the atmosphere.  Ans. As 61 : 48.

7. 100 cubic centimetres of air at 0° C are heated to 300° C under constant pressure, what will be the volume of the air at the higher temperature if the coefficient of expansion = ·00366 ?  *U. of L. Matriculation*, 1874.

8. If the coefficient of expansion of atmospheric air for the centigrade scale be ₂₇₃, find the temperature to which 500 cubic centimetres of air (measured at 15° C) must be raised in order that its volume may become 700 cubic centimetres, no change of pressure taking place meanwhile.

*U. of L. Matriculation*, 1875.

192. Knowing the coefficient of expansion for air, we can conversely find the temperature imparted to it by any source of heat, if we find experimentally the expansion given to any known volume of air. An instrument employed for this purpose is called an *air thermometer*. It consists simply of a retort-shaped vessel of glass or porcelain with its tube of a narrow bore, and drawn out at its extremity to a very fine point. If such an instrument be placed in any body whose temperature is to be determined, as for instance in a furnace, the enclosed air will be expanded, and be driven out through the tube, until its temperature has become the same as that of the furnace. When this point is reached the tube is hermetically sealed, and the vessel is withdrawn from the furnace and gradually cooled. When at 0° C the point of the tube is broken off under mercury, and the mercury will enter the vessel until the density of the contained air is equal to that due to the atmospheric pressure. The weight of the mercury is then found, and also the weight of mercury which just fills the vessel. The ratio of the difference of these weights to the latter weight is the ratio of the volume of the contained air at 0° to its volume at the temperature of the furnace; that is, if *t* be temperature of the furnace, it

is as 1 to $1 + \dfrac{t}{273}$ or as 273 to $273 + t$. Whence the value of $t$ is readily found. For instance, if the weight of mercury which enters the vessel when cooled to 0° C is 7000 grammes, and the vessel holds 10000 grammes of mercury, then

$$3000 : 10000 :: 273 : 273 + t$$
$$3t = 273 \times 7$$
$$t = 637°\text{C}.$$

**193. Fusion or liquefaction.**—A change of volume is not the only phenomenon which results from a change in the temperature of a body. Through a change of temperature a body may undergo a change of state—may pass from a solid into a liquid, or from a liquid into a gaseous state, or *vice versâ*. *Fusion* is the term used to denote the change which takes place when a body passes from the solid into the liquid state. The laws of fusion are these :

1. *Every substance begins to fuse at a certain temperature, which is constant for each substance if the pressure be constant.*

2. *From the moment when fusion commences, the temperature of the substance remains at this constant point until the fusion is complete.*

So, conversely, when a body undergoes solidification, or passes from the liquid into the solid state, similar laws obtain.

1. *Every substance solidifies at a certain temperature, which is constant for each substance if the pressure be constant.*

2. *From the moment that solidification commences, the temperature of the substance remains at this constant point until the solidification is complete.*

In passing from the liquid to the solid state some substances (such as water, cast-iron, bismuth, and antimony) expand, and some (as mercury, gold, silver, copper, and probably most substances) contract. In reference to this peculiarity the following additional law is found to obtain :

3. *If a substance expand in solidification, its melting-point is lowered by pressure ; but if it contract, the melting-point is raised by pressure.*

An illustration of this law is supplied by the motion of glaciers. In consequence of the great pressures to which they are subjected, the lower portions, notwithstanding the lowness of their temperature, are reduced to a viscous or semi-liquid state, and the glacier is thus able to move down valleys and ravines without any violent rupture of its substance.

**194. Fusing point.**—The following table exhibits the points of fusion of different substances under the average atmospheric pressure (760 mm.):

| | | | | | |
|---|---|---|---|---|---|
| Mercury | . . . | 39° C | Tin | . . . . | 233° C |
| Ice | . . . . | 0° | Lead | . . . . | 327° |
| Phosphorus | . . | 44° | Zinc | . . . . | 360° |
| Potassium | . . | 58° | Silver | . . . | 1000° |
| Sodium | . . . | 98° | Gold | . . . . | 1250° |
| Sulphur | . . . | 115° | Iron | . . . . . | 1500° |

The fusing-point is also the point of congelation, or the temperature at which a body passes from the liquid into the solid state.

**195. Alloys, fluxes.**—The fusing of a mixture is often found to be lower than that of either of the substances of which it is composed. An alloy of five parts of tin and one of lead fuses at 194° C. Rose's fusible metal, consisting of four parts of bismuth (fusing-point 267° C), one of lead, and one of tin, fuses at 94° C. A substance added to another in order to facilitate the fusion of the latter is called a *flux*.

**196. Vaporization** is a general term used to denote the change which takes place when a body passes from a liquid into a gaseous state, under whatever circumstances that may occur. It includes both ebullition and evaporation.

*Ebullition* takes place when the change of a liquid into the gaseous state is attended by the formation of bubbles of vapour in the mass of the liquid itself.*

*Evaporation* takes place when a liquid passes into a gas without any formation of bubbles within the mass of the liquid, the vapour being formed at the surface only.

**197. Laws of ebullition.**—1. *All liquids begin to boil at a temperature which for each liquid is constant under a constant pressure.*

2. *The moment ebullition commences the liquid remains at this constant temperature until it is entirely vaporised.*

3. *As the pressure increases the temperature of the boiling-point increases.*

* Most liquids contain more or less of air in solution. When heated the bubbles which first arise are bubbles of air, and not of vaporised liquid. It is not until all the air is disengaged, and bubbles of vapour begin to rise, that ebullition is said to commence.

**198. Boiling point.**—The temperature at which different substances begin to boil is very varied. The following table gives the boiling-point of the substances named under the ordinary atmospheric pressure (Barometer at 760 mm.):

| | | | |
|---|---|---|---|
| Sulphurous Anhydride | 8° C | Benzole . . . | 80° C |
| Ether . . . . . | 35° | Water . . . | 100° |
| Bisulphide of Carbon . | 48° | Fusel Oil . . | 132° |
| Bromine . . . . . | 63° | Sulphuric Acid . | 338° |
| Alcohol . . . . . | 78° | Mercury . . . | 350° |

**199.** *When a liquid boils under any pressure the tension of the vapour is equal to that pressure.*

The pressure exerted by any vapour, or the elastic force of the vapour, is commonly spoken of as its *tension;* and the principle just enunciated may be experimentally verified by the following method : AB is a barometer tube, surrounded by a jacket, which at its upper end E is connected 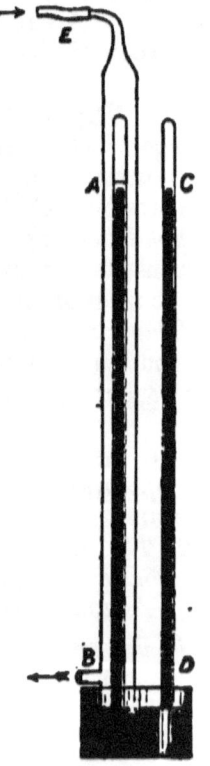 with a flask, in which water or any other liquid can be boiled, and at its lower end has a small opening, through which the vapour may escape. A small quantity of liquid, say water, is introduced into the Torricellian vacuum of the enclosed barometer AB, and aqueous vapour is formed therein. Steam from the flask passing round AB gradually raises the temperature of this vapour. As the temperature of the vapour rises the mercury falls in AB, and when the vapour has reached the temperature of the surrounding steam, the mercury in AB is at the level of the mercury in the trough, and at this point it remains so long as this temperature is maintained. The tension of aqueous vapour at the temperature of boiling water is thus seen to be equal to the atmospheric pressure.

A similar result is obtained if, instead of water, ether be introduced into the Torricellian vacuum of AB, and the vapour of boiling ether be admitted into the jacket from the flask. So also with any other liquid ; and hence when a liquid boils under the pressure of the atmosphere, the tension of its vapour is equal to the atmospheric pressure.

Since under the same atmospheric pressure different liquids boil at different temperatures, it follows that different vapours have the same tension at different temperatures ; and, conversely, that at the same temperature different vapours have different tensions, the tension being the greater according as the boiling-point is the lower.

200. By means of the apparatus described in the preceding article the tension of aqueous (or any other) vapour at temperatures lower than the boiling-point may be found. Let water be introduced into the Torricellian vacuum of AB as before, and let the vapour of some liquid that boils at a lower temperature than water be introduced into the jacket from the flask. The aqueous vapour in AB will then ultimately be raised to the temperature of the surrounding vapour, and its tension will be found by taking the difference between the height of the mercury in AB and the height of the mercury in the free barometer CD. Thus if the latter stand at 760 mm., and the former at 434 mm., the tension of the aqueous vapour in AB is seen to be equal to the weight of 326 mm. of mercury.

201. Since the tension of the vapour when a liquid boils is equal to that of the surrounding pressure, and since also the tension of a vapour increases with its temperature, it follows that if the pressure be diminished or increased, the temperature at which boiling takes place will also be diminished or increased.

Hence at great altitudes, where the atmospheric pressure is less than at the surface of the earth, the temperature of boiling water is considerably lower than 100° C, and consequently is often insufficient for culinary purposes. On the top of Mont Blanc water boils at 84° C.

The observation of the temperature at which water boils is thus one method by which the height of mountains may be determined.

The ebullition of water at a low temperature under diminished pressure may be exhibited by an easy experiment. Take a common flask and partly fill it with water. Boil this water over a lamp, and when the upper part is filled with vapour remove the flask from the lamp, and close it tightly with a cork. The water will then cease to boil, and will begin to fall in temperature. If now a sponge moistened with cold water be applied to the upper part of the flask ebullition will recommence. The application of the sponge has partly condensed the vapour, and consequently diminished the pressure it exerts upon the water, and at this reduced pressure the temperature of the water is sufficient to cause it to boil.

If a glass containing ether be placed under the receiver of an air-pump, the ether will boil at ordinary temperatures when the receiver has been partially exhausted.

202. **Laws of evaporation.**—As already stated, evaporation is the production of vapour at the surface of a fluid. It is the process which takes place when articles are dried by exposure to the air, or when a liquid gradually disappears from an open vessel. The following laws of evaporation have been deduced from experiments:

1. *Evaporation takes place the more rapidly the greater the extent of surface exposed.*

2. *It increases with the increase of temperature.*

3. *It decreases as the quantity of the same vapour in the surrounding atmosphere increases.*

Air is capable of holding only a certain quantity of any vapour in solution, and consequently when the surrounding air is saturated with the vapour of any liquid no further evaporation from that liquid will take place. Hence,

4. *Evaporation is facilitated by the agitation of the surrounding air, or by any other cause which supplies a drier air in the place of that which has become wholly or partially saturated with the vapour.*

203. **Latent heat.**—It has been seen that while a body is passing from the solid to the liquid state the temperature remains constant until the fusion is complete, however great the intensity of the heat applied. It hence follows that a portion of the heat is employed in effecting the change of state. This is termed the *latent heat of fusion.* Similarly the heat employed in effecting the change from the liquid to the gaseous state is termed the *latent heat of vaporization.*

If 1 lb. of water at 79° C be mixed with 1 lb. of ice at 0° C, there will result 2 lbs. of water at 0°. The quantity of heat lost by the former of the two ingredients is 79 thermal units, and as the temperature of the other ingredient has not been raised, it follows that this quantity of heat has been wholly spent in melting the ice. The quantity of heat required to melt one part by weight of ice at 0° is thus seen to be 79 times as great as that required to raise one part of water 1° C, and hence using the centigrade scale the latent heat of fusion for water is said to be 79. (On the F scale it is 142.)

If 1 lb. of steam at 100° C be mixed with 9 lbs. of water at 0°, the resulting temperature will be found to be about 63°·6 C. Here the amount of sensible heat before mixture is 100 thermal units. After mixture we have 10 × 63·6, or 636 thermal units. The difference of these, or 536 thermal units, is consequently the amount of heat given out by 1 lb. of steam in passing into water, which is the same as the amount of heat absorbed by 1 lb. of water in passing into steam. The latent heat of steam therefore is approximately 536. (On the F scale, 965.)

Methods of finding the latent heat of other substances will be given further on.

204. The following tables exhibit some of the results obtained by careful experiments :

### LATENT HEAT OF FUSION.

| Water | 79·25 | Bismuth | 12·64 |
|---|---|---|---|
| Nitrate of potash | 62·98 | Sulphur | 9·37 |
| Zinc | 28·13 | Lead | 5·37 |
| Silver | 21·07 | Phosphorus | 5·03 |
| Tin | 14·25 | Mercury | 2·83 |

### LATENT HEAT OF VAPORIZATION.

| Water | 535·9 | Acetic acid | 102 |
|---|---|---|---|
| Wood spirit | 263·7 | Ether | 90·5 |
| Alcohol | 202·4 | Bisulphide of carbon | 86·7 |
| Formic acid | 169 | Oil of turpentine | 69 |

As by observing the temperature of the result when ice is melted by mixture with water we learn the latent heat of water, so conversely knowing the latent heat of water we can calculate the resulting temperature when a given quantity of ice is mixed with a given quantity of water at a known temperature ; for the total quantity of sensible heat after mixture will be equal to the sensible heat before mixture, minus the heat absorbed in the fusion of the ice.

*Ex.* Four parts by weight of ice at o° C are mixed with six parts of water at 100°, required the temperature when the ice has been wholly melted.

Let *x* be the required temperature, then 10*x* will be the quantity of sensible heat after mixture. The quantity of sensible heat before mixture is 6 × 100 or 600. The heat absorbed in melting the ice is 4 × 79 or 316. Hence,

$$10x = 600 - 316 = 284$$
$$\therefore \quad x = 28°·4 \text{ C.}$$

### EXAMPLES.

(Latent heat of water = 79.  Latent heat of steam = 536.)

1. If 2 lbs. of ice at o° C be mixed with 5 lbs. of water at 61° C, what will be the temperature of the mixture ? *       Ans. 21° C.

2. What weight of water at 100° C will be necessary just to melt 10 lbs. of ice at o° ?       Ans. 7·9 lbs.

3. How much ice at o° is necessary to cool 1 lb. of water at 100° C to o° C ?       Ans. 1·26 lbs.

4. How much ice at o° must be mixed with 9 lbs. of water at 20° C to cool it to 5° ?       . Ans. 1·607 lbs.

5. What weight of steam at 100° C is necessary to raise 200 lbs. of water from 18° C to 36° C ?       Ans. 6 lbs.

6. If 4 kilogrammes of snow at - 10° C be mixed with 8 kilogrammes of water at 60°, what is the resulting temperature ?       Ans. 2° C.

7. If in the preceding the proportions of snow and water were reversed, show that the result would be nearly equal portions of snow and water at o° C.

8. The heat produced by the complete combustion of one gramme of carbon in a calorimeter can convert 100 grammes of ice at o° C into water at o° C. How many grammes of water could be raised by the same amount of heat from o° C to 1° C ?       *U. of L. Matriculation, 1873.*

9. How much ice at o° C can be converted into water at o° C by an ounce of steam at 100° C if we assume heat to be transmitted from the steam only to the ice ?  (Latent heat of water, 80; latent heat of steam, 536.)  *U. of L. Matriculation, 1875.*

10. It is found that a kilogramme of water at 100° C, mixed with a kilogramme of melting snow without loss of heat, gives two kilogrammes of water at the temperature of 10°·3 C.  Show how to find from this the latent heat of water.  *U. of L. Matriculation, 1875.*

**205. Specific heat.**—The quantity of heat required to raise the temperature of any body through any given number of degrees is not the same for all substances.

* In this and the following examples the mixture is supposed to take place without any loss of heat.

For instance, if equal weights of water at 100° C and water at 0° C be mixed, the resulting temperature is 50° C; but if equal weights of mercury at 100° and water at 0° be mixed, the resulting temperature is only 3° C; while if equal weights of water at 100° and of mercury at 0° be mixed, the resulting temperature is 97° C. Hence the heat which raises or lowers water 3° C lowers or raises the same weight of mercury 97° C.

*The specific heat of any substance is the ratio which the quantity of heat necessary to raise its temperature 1° bears to the quantity of heat necessary to raise the same weight of water 1°.*

It follows from this definition that the specific heat of any substance is the number of thermal units required to raise a weight-unit of that substance 1° in temperature. And hence if $c$ be the specific heat of any substance, $m$ its weight, and $t$ its temperature (reckoned from freezing point), $cmt$ is the quantity of heat taken in, in passing from 0° to $t°$, or given out, in passing from $t°$ to 0°; in other words, $cmt$ represents the quantity of sensible heat in the substance.

206. If two substances be mixed without any change of state, care being taken to prevent any loss of heat during the operation, the total quantity of sensible heat after mixture is the same as before. Hence the temperature of any such mixture may be calculated if the temperatures, weights, and specific heats of the components are known.

Thus if 3 lbs. of A, whose temperature is 100° C and specific heat ·03, be mixed with 5 lbs. of B, whose temperature is 40° C and specific heat ·08, the temperature of the mixture may be thus found. Let $x$ be the required temperature.

After mixture the quantity of heat in A is $·03 \times 3 \times x$, and the quantity of heat in B is $·08 \times 5 \times x$.

Before mixture the quantity of heat in A is $·03 \times 3 \times 100$, and in B $·08 \times 5 \times 40$; hence,

$$·09x + ·4x = 9 + 16$$

$$x = \frac{25}{·49} = 51°·02 \; C.$$

207. The preceding example suggests a method by which the specific heat of any substance may be conveniently found. For if the specific heat of one of two components be known, and the temperature of the mixture be found by experiment, the specific heat of the other component may be determined by equating the quantities of heat before and after mixture. Water

M

being the standard of comparison, its specific heat is 1. Hence to find the specific heat of any substance mix any quantity of known weight and temperature with water of known weight and temperature, and observe the temperature of the result. From these data the specific heat may be determined, as in the following example:

2 lbs. of iron heated to 100° C are placed in 4 lbs. of water at 30° C, the resulting temperature is found to be 33·9° C, it is required to find the specific heat of iron.

Let $x$ be the specific heat of the iron, then

quantity of heat after mixture $= x \times 2 \times 33\text{·}9 + 4 \times 33\text{·}9$

„        before   „    $= x \times 2 \times 100 + 4 \times 30$

$67\text{·}8x + 135\text{·}6 = 200x + 120$

$132\text{·}2x = 15\text{·}6$ or $x = \text{·}118.$

This method of finding the specific heat of a substance is generally described as the *method of mixture.*

208. The specific heat of a substance may also be found by the fusion of ice. A ball or lump of the substance heated to $t°$ C is placed in a cavity formed in a block of ice until it has cooled down to 0° C. A portion of the ice will consequently be melted into water. This water is then carefully weighed, and its weight multiplied by 79 (the latent heat of water) is the quantity of heat given out by the body in cooling from $t°$ to 0°; that is, the quantity of heat required to raise the temperature from 0° to $t°$. This divided by $t$ and by the weight of the body will give its specific heat.

A block of ice prepared for this purpose, with a slab of ice covering the cavity, is called Black's, or the Ice-block *Calorimeter.*

209. **Table of specific heats.**—The following table exhibits some of the results obtained by careful experiments made in accordance with the two methods described above, and give the mean specific heat for temperatures between 0° C and 100° C :

| | | | | |
|---|---|---|---|---|
| Water . | . | 1 | Iron . . | 0·1138 |
| Alcohol | . | 0·6735 | Zinc . . | 0·0955 |
| Acetic Acid . | | 0·6589 | Copper . | 0·0951 |
| Ether . . | | 0·5157 | Silver . . | 0·0570 |
| Turpentine . | | 0·4629 | Tin . . | 0·0562 |
| Wood Charcoal | . | 0·2415 | Gold . . | 0·0324 |
| Glass . . | | 0·1770 | Lead . . | 0·0314 |

210. The method of mixture described in Article 206 may be also used for finding the *latent* heat of any substance. If the quantity of sensible heat after mixture be greater or less than the quantity before mixture, latent heat has been disengaged or absorbed through a change of state in one or more of the ingredients. By observing this difference when one only of the ingredients undergoes a change of state, we obtain the quantity of latent heat absorbed or disengaged by that ingredient, and this divided by its weight will give the measure of its latent heat.

Thus, let three parts by weight of molten lead at 400° be placed in one part of water at 0°, and let the resulting temperature be 49° C.

The specific heat of lead is ·0314; hence the quantity of sensible heat before mixture is ·0314 × 3 × 400 = 37·68.

After mixture the sensible heat in the lead is ·0314 × 3 × 49 = 4·62 nearly, and in the water 49, or 53·62 in all.

The difference, 53·62 − 37·68 or 15·94, is the latent heat given by the solidification of the lead. This divided by 3, the weight of the lead, gives 5·31 as approximately the latent heat of this metal.

### EXAMPLES.

1. A piece of metal weighing 600 grms. is heated to 100° C and plunged into 2000 grms. of water at 0° C ; the resulting temperature is 3° C, what is the specific heat of the metal ? *Ans.* ·103.

2. A piece of glass (specific heat = ·177) weighing 60 grms. is heated to 80° C and plunged into 400 grms. of a liquid at 6° C ; the resulting temperature is 9° C, what is the specific heat of the liquid ? *Ans.* ·628.

3. If 10 oz. of mercury at 100° C be agitated with 1 oz. of water at 10° C, what will be the temperature of the mixture, assuming the specific heat of mercury to be ·033 ? *Ans.* 32°·33 C.

4. If 10 parts by weight of alcohol at 20° C be mixed with 2 parts of water at 10° C, what will be the temperature of the mixture ? *Ans.* 17°·71 C nearly.

5. How much water at 0° C must be poured upon 2 lbs. of iron at 250° C to cool the iron down to 20° C ? *Ans.* 2·5036 lbs.

6. What weight of iron at 500° C must be placed in 4 lbs. of water at 0° to raise it to 20° C ? *Ans.* 1·46 lbs.

7. How many pounds of molten lead at 327° C will just melt 1 lb. of ice at 0° C ? *Ans.* 5·174.

8. If equal weights of two substances at different temperatures have, when mixed, a resulting temperature which is a mean between the two, show that the specific heats of the substances are equal.

9. If the weights of two substances of different temperatures are inversely as their specific heats, then the resultant temperature of their mixture is the mean of the two temperatures.

10. What must be the relation between the specific heat and the specific gravity of any substance, that when mixed with an equal volume of water at a different temperature the resultant temperature may be the mean of the two temperatures?

Ans. The sp. heat is the reciprocal of the sp. gravity.

11. If the specific heats of two liquids are inversely as their specific gravities, show that equal volumes of the liquids at different temperatures will after mixture have a temperature which is the mean of the two temperatures.

**211. Transmission of heat.**—There are three modes in which heat is transmitted from one point to another in space; (1) by conduction, (2) by convection, and (3) by radiation.

Heat is transmitted by *conduction* when it passes from one point to another in the same body without any sensible change in the relative position of the particles of the body.

Heat is transmitted by *convection* when it is diffused through a fluid body by the motion of its heated particles.

Heat is transmitted by *radiation* when it passes from a heated body to any point in free space, or if through any medium, without affecting the temperature of that medium.

Although it is convenient to consider these modes of transmission apart from one another, it is important to observe that under ordinary circumstances diffusion of heat takes place in all three modes combined. Thus if a metal rod be placed with one end in the fire, heat is transmitted along the rod, and so far there is the conduction of heat. On reaching the surface of the rod, the heat is partly communicated *to* the air immediately in contact with it, and partly is diffused *through* the air by radiation. The air that is heated by contact with the rod rises, through its diminished density, and is replaced by cooler air, which in its turn becomes heated and rises, and so some part of the heat is diffused through the surrounding air; and this is the process which is described by the term convection.

**212. Conduction of heat in solids.**—If one end A of a metal or other rod be maintained at a constant temperature,* and the temperature of any point B, at a short distance from A, be tested by a thermometer, it will be found that the temperature at B will gradually rise until it has attained a certain

---

* This may conveniently be done by inserting the rod in the side of a vessel containing a heated liquid, say water retained at boiling heat; or, if a higher temperature be required, molten lead, or some other metal maintained at its point of fusion.

maximum, less than the temperature at A; and at this maximum
it will remain as long as the constant temperature is maintained
at A. It will further be found that as the distance of B from
A is increased, the lower is this maximum temperature, until at
a point C the temperature of the bar is simply that of the surrounding air; that is to say, no heat at all is transmitted from
A to C, and therefore also to no point beyond C.

If different rods be similarly tested, similar results obtain;
but the distance of C from A is different in different substances,
and that substance is said to have the greatest conductivity in
which this distance is the greatest.

Hence if similar rods of different substances have each an
extremity maintained at the same constant temperature, the
temperature at points equidistant from this extremity will be
the greater according as the conductivity of the rod is the
greater. And also the greater the conductivity, the more distant
from the extremity will be the points of equal temperature.

If then a series of similar rods formed of different substances
be coated with wax, and inserted in the sides of a vessel containing water at 100° C, the different conductivity of the rods
will be shown by the different lengths from which the wax will
be melted.

213. Of solid substances, metals are generally speaking good
conductors; organic substances bad conductors; and substances
of loose texture, such as wool, feathers, bran, conduct worst of
all. Of metals, some have much greater conductivity than
others. Taking 100 to denote the conductivity of silver, the
following numbers are given as denoting the relative conductivity
of other well-known metals:

| | | | | | | |
|---|---|---|---|---|---|---|
| Copper | . | . | . 73·6 | Steel | . . | . 11·6 |
| Gold | . | . | . 53·2 | Lead | . . | . 8·5 |
| Tin . | . | . | . 14·5 | Platinum | . | . 8·4 |
| Iron | . | . | . 11·9 | Bismuth . | . | . 1·8 |

NOTE.—The student must guard against assuming that the greater
the conductivity, the greater also the velocity of transmission. Although
it may be that of two substances, the one which has the greater conductivity transmits the more quickly, this is not invariably so.

214. **Safety lamp.**—If a piece of wire gauze be placed in
the body of a flame, the flame will not pass through the gauze,
the reason being that the heat is so largely withdrawn by the

wire, that the temperature is lowered beyond the point at which
the gas is ignited.   In like manner, if the gauze be held over a
jet of unignited gas, and the gas above the gauze be ignited, the
flame will not pass through the gauze to the gas below.

The safety lamp is an ingenious application of this fact made
by Sir Humphrey Davy for the protection of miners.   It consists
of an ordinary lamp enclosed in wire gauze.   If a dangerous
explosive gas arise in the mine, a portion of it passes through
the gauze and is exploded by the flame within, thus giving
warning to the miner of the presence of the gas; but, except
when the explosion is unusually violent, the flame within the
lamp does not pass through the gauze, and the exterior gas is
unaffected by it.

**215. Conduction of heat in liquids and gases.**—The
conductive power of liquids is very small, and that of a gas is
still less.   If a vessel containing heated oil be placed in contact
with the surface of some water, it is found that the layers of
water near the surface are affected in their temperature in but a
very small degree, and this becomes sensible only after the lapse
of a considerable time.   Similar results obtain generally in the
case of other liquids, mercury and other fused metals excepted.

So slight is the conductive power of water that if a piece of
ice be fixed in the lower part of a tube partly filled with water,
heat may be applied to the middle of the tube so as to cause
the upper part of the water to boil and the ice at the bottom
shall remain unaffected.

It is owing to the inconsiderable conductivity of air that furs,
woollen cloth, and other similar substances are useful as articles
of clothing.   Being bad conductors of heat because of the layers
of air enclosed within them, they prevent the heat of the body
from rapidly passing away.   For a similar reason double windows
are useful in maintaining the warmth of houses in cold weather.

**216. Convection of heat in fluids.**—As already explained,
the diffusion of heat through a fluid mass arises from the motion
from the lower to the higher part of the mass of the portions
which successively become heated through the application of
flame or other source of heat.   When heat is applied at the
bottom of any vessel containing a liquid, the heated portion
becoming specifically lighter than the surrounding liquid rises
to the surface, and the cooler liquid descends into its place.   A

double current is thus occasioned, one moving upward and the other downward, the upper current being in the central part of the vessel and the downward current at the sides. If the lower part of the liquid be coloured, or if some small substances such as the sawdust of heavy wood be thrown in, these currents may be made distinctly apparent.

217. **Radiation of heat.**—That heat passes off like light in rectilinear directions from every point of a heated surface may be shown by some simple experiments.

If we stand in front of a fire with our face towards it we are sensible of an increase of heat, but if with our face away from it no such increase is felt. The sensation therefore does not arise from the heat imparted by the fire to the air in the room.

Again, if we allow the heat from a fire or a red-hot ball to fall directly upon a thermometer, the mercury will indicate a rise of temperature; but if a screen be interposed between the thermometer and the source of heat the thermometer is unaffected.

That the passage of heat is rectilinear can be shown as in the case of light. If three screens, each of which has a small opening, be so placed that the three openings be in the same straight line, a heated ball placed in front of the opening in the first screen will affect a thermometer placed behind the opening in the third screen; while if the three openings be not in the same straight line the thermometer will be unaffected.

218. **Reflection of heat.**—Like light, heat is reflected from smooth surfaces, and according to the same laws; viz., 1, *The reflected ray is in the same plane with the incident ray and the normal at the point of incidence;* and 2, *the angle of reflection is equal to the angle of incidence.*

This may be verified by showing experimentally that the necessary consequences of these laws, already deduced for light, apply also in the case of heat. Thus, if two concave mirrors be placed facing each other, with their axes in the same straight line, and a heated ball be placed in the focus of the one mirror and a thermometer in that of the other, the rise of the mercury will show the concentration of the heat in that spot. That this has not arisen from the heat passing directly from the ball to the thermometer may be shown by moving the thermometer out of the focus, for even though it be moved nearer to the ball the thermometer will fall. The direct heat may even be cut off by a small screen of flannel, and the thermometer when in the focus will still be affected. The rays of heat falling upon the first mirror have been reflected parallel to the axes (see fig. Art. 149), and these parallel rays falling upon the second mirror have been reflected to its focus, and so have affected the thermometer placed there.

In like manner, if the heated ball be placed in one of the conjugate foci of a concave mirror ($p$ fig. Art. 152), and the thermometer in the other (P fig. Art. 152), the rays from the ball will converge thereupon, and a considerable rise of temperature will be indicated.

219. **Refraction of heat.**—Like light, heat may also be refracted. As some substances allow light to pass readily through them, and are called transparent, so some substances allow heat to pass through them, and are called *diathermanous*. Rock salt is such a substance; and if a lens be made of rock salt, heated rays will be found to be affected by it precisely in the same way as light rays are affected by a glass lens.

220. In other respects also the properties and laws of radiant heat are similar to those of light. These are not sufficiently elementary to be explained here; it may however be mentioned that they all lead up to the conclusion that heat and light are in all probability modifications of the same force, and that hence the velocity of transmission is the same for radiant heat as for light, or at the rate of 186,000 miles per second.

# CHAPTER X.

221. When the particles of an elastic body are suddenly displaced at any point, a peculiar motion termed *vibration* is communicated to all the particles of that body.. The displaced particle oscillates or vibrates through small distances about its position of equilibrium, its excursions on either side gradually diminishing, and at length altogether ceasing. A similar vibratory movement is communicated to the other particles of the body. The velocity with which a vibration is transmitted from one point in the body to another is different in different bodies, and must carefully be distinguished from the absolute velocity of the particles themselves.

An elastic body, in a state of vibration, may communicate a similar motion to the particles of another elastic body in contact with it.

222. **The sensation of sound produced by the vibration of an elastic body in contact with the ear.**—If air be the medium in contact with the ear, as is most commonly the case, sound is produced by the vibrations of the air; if water be the medium in contact with the ear, the vibrations of the water produce the sensation; if an iron or glass rod, then the vibrations of the particles of the rod produce it.

The vibratory motion which produces the sensation of sound is either produced directly in the medium in contact with the ear, or is communicated to that medium by the vibration of another elastic body. When a bell is struck, a vibratory motion is given to the particles of the bell, and we hear the sound, because vibrations are thence communicated to the surrounding air, which is the medium in contact with our ear. If, however, the bell be so placed that its vibrations cannot be communicated to the air, no sound will be heard, however violently it may be rung. For instance, if the bell be suspended under the receiver of an air pump, and be made to ring, as rarefaction proceeds the sound becomes gradually weakened, and at length becomes inaudible.

223. **The velocity of sound.**—The average velocity of the propagation of sound in atmospheric air is found by experiment to be about 1130 feet in a second. In water, the velocity of sound is 4900 feet in a second, and along a deal rod the velocity is as great as 17,000 feet in a second.

Increase of temperature increases the velocity of sound. According to the best experiments, the velocity of sound in air at 62° Fahrenheit is 1125 feet in a second. Every increase or decrease of temperature of 1° Fahrenheit causes a corresponding increase or decrease of 1·14 feet in the velocity of sound. Accordingly, the velocity of sound in air at the freezing-point will be 1090 feet per second.

The velocity of sound in a gas is found to be affected by a change in its chemical nature. According to experiments made by M. Dulong, the velocity of sound at the freezing-point was, in

| | | |
|---|---|---|
| Carbonic Acid . . . | 858 feet per second. | |
| Protoxide of Nitrogen . | 859 ,, | ,, |
| Oxygen . . . . | 1040 ., | ,, |
| Carbonic Oxide . . | 1105 ,, | ,, |
| Hydrogen . . . | 4165 ,, | ,, |

It will be hence seen that increase of density does not necessarily produce an increase in the velocity of sound. Carbonic acid, which is more dense than atmospheric air, transmits sound with a less velocity; while hydrogen, which is rarer than atmospheric air, transmits sound with a velocity nearly four times as great.

In any given medium, the velocity of propagation is *in general* found to be the same for all sounds, whatever their loudness or tone. In experiments made by M. Biot, several airs played on a flute, at one end of an iron tube 1040 yards in length, were distinctly heard by him at the other end, without any variation whatever; and a case is recorded by Mr. Besant, in which the Hallelujah Chorus, played by a band on a fine and still evening, was heard, without any loss of harmony, at a distance of two miles.  As far as these experiments go, they shew that the velocity of different sounds in the same medium is the same, or at least does not sensibly vary.  This, however, is not *universally* true.  Recent experiments have shewn, that two sounds differing very greatly in intensity travel with sensibly different velocities, and that a very loud sound travels faster than a feeble one.

**224. Phenomena explained by the velocity of sound.** —It follows from what has been stated respecting the velocity of sound, that sounds, simultaneously produced at a series of points, will reach the ear in the same time, if the points are equidistant from the ear; but at different times, if the points are at different distances.  This furnishes an explanation of several phenomena, such as, for instance, the crash and rolling of thunder.  The velocity of lightning is so great, that the sounds produced at the various points of a flash may be regarded as simultaneously produced.  If then the track of the lightning be across the line of the observer, so that its various points are equally distant, or nearly equally distant, from him, all the sound will reach his ear at the same, or nearly the same instant, and the result will be an explosive sound of great intensity.  If, on the other hand, the path of the lightning is towards, or from the observer, then the different points in it being at different distances from him, the sound will reach him in different times, and consequently the sensation produced will be that of a continuous or rolling sound.  Sir John Herschel illustrates it thus: "Conceive two equal flashes of lightning, each four miles long, both beginning at points equidistant from the auditor; but the one running out in a straight line directly away from him, the other describing an arc of a circle, having him in its centre.  The thunder may be regarded as originating at one and the same instant at every point of either flash; but it will reach the ear under very different circumstances in the two cases.  In that of the circular flash, the sound from every point will arrive at the same instant, and affect the ear as a single explosion of stunning loudness.  In that of the rectilinear flash, on the other hand, the sound from the nearest point will arrive sooner than from those at a greater distance; and those from different points will arrive in succession, occupying altogether a time equal to that required by sound to run over four miles, or about 20 seconds.  Thus the same *amount* of sound is in the latter case distributed uniformly over 20 seconds of time, which in the former arrives at a single burst; of course it will have the effect of a long roar, diminishing in intensity as it comes from a greater and greater distance. If the flash be inclined in direction, the sound will reach the ear *more compactly* (*i.e.* in a shorter time from its commencement) and be proportionally more intense. If (as is almost always the case) the flash be zigzag, and composed of broken rectilinear and curvilinear portions, some concave, some convex to the ear; and if, especially, the principal trunk separates into many branches, each breaking its own way through the air, and each becoming a separate source of thunder, all the varieties of that awful sound are easily accounted for."*

A similar explanation applies to the difference observed in the sound produced by the discharge of a long file of musketry when the hearer stands directly in front or behind the file, at some distance from it, and when he stands in a line with it. In the former case, the sound reaching him from all parts of the file at nearly the same instant, a sharp loud report is produced. In the latter case, the different parts of the file being at different distances, the sound produced is continuous, like that of a running fire.

225. **Qualities of sound.**—Sounds are divided into two principal classes, viz., *musical* and *non-musical* sounds; or, in other words, into *notes* and *noises*. If a series of vibrations, performed in equal times, strike upon the ear, a pleasing effect is produced, and the sound is distinguished as a musical sound or note; but when the vibrations are performed in unequal times, the impression made is less pleasing, and the sound is non-musical, or noise.

Any two sounds of either class may differ from each other in the following respects; first, in *intensity;* secondly, in *pitch;* and thirdly, in *timbre* or that peculiarity by which the sounds produced by one instrument are distinguished from those produced by another.

226. **Intensity of sound.**—The intensity of sound regarded as a sensation is, like all other sensations, largely dependent upon the sensitiveness of the organ by which it is perceived. The ear of one person may be more sensitive to sound than that of another, and the ear of the same person may be more

* Encyclop. Metrop., Art. Sound, 39.

sensitive at one time than at another time. The consideration of this cause of variation in the intensity of sound belongs to the science of physiology. The following are those causes which fall within the province of acoustics. The intensity of sound depends,

1st. Upon the greatest absolute velocity of the vibrating particles in the sounding body; or, which amounts to the same thing,* upon the extent, or *amplitude*, as it is termed, of the vibrations. All other things being equal, the greater the absolute velocity of the particles, and consequently the greater the amplitude of the vibrations, the greater is the intensity of the sound. This is shewn by common experience; for the harder any sounding body is struck, a drum or a bell for instance, the louder is the sound produced.

2ndly. Upon the distance of the hearer from the origin of the sound. The greater the distance the less the intensity of the sound. If the sound be transmitted through a mass of free air, that is, of air not confined by any pipe, or not bounded by any reflecting surface, the intensity of the sound varies inversely as the square of the distance. Thus, at double the distance, the intensity is one-fourth, at treble the distance, it is one-ninth, and so on. When, however, sound is transmitted through the air contained in a tube of small bore, the intensity of the sound is but very slightly diminished, even at great distances. In experiments made by M. Biot, upon a tube 1040 yards in length, formed by the cast-iron water pipes of Paris, the slightest whisper made at one end of the tube was distinctly heard at the other. It is from this property of transmitting sounds that speaking tubes derive their value.

3rdly. Upon the position of the hearer in relation to the original direction of the sound. When the hearer is in a line with the direction of original sound, the intensity is greater than when the sound is transmitted in a direction inclined to this line. If a tuning-fork be struck and held at a short distance from the ear, and afterwards be turned round on its axis, the sound is perceptibly louder when the flat side is towards the ear than when the flat side is turned away from the ear.

4thly. Upon the density of the atmosphere (or other medium) through which the sound is transmitted. We have already seen

* For clearly the greater the velocity impressed upon a particle, the greater the distance it will move from its position of rest.

that when a bell is placed under the receiver of an air pump, the intensity of the sound diminishes as the air in the receiver becomes more rare. On the top of high mountains, where the air is considerably rarified, it requires a great effort to be heard at a moderate distance, and the firing of a pistol produces only a comparatively slight report. On the other hand, in the condensed air of a diving bell, the ordinary effort in speaking produces sounds which are unpleasantly loud ; and in the experiments made at the lake of Geneva by M. Colladon, in the year 1826, the sound of a bell struck under the water was distinctly heard across the whole breadth of the lake from Rolle to Thonon, a distance of about 9 miles.

Sounds produced in oxygen and carbonic acid are more intense, and in hydrogen are more feeble, than those excited by the same cause in atmospheric air. Oxygen and carbonic acid are both denser, and hydrogen is much rarer, than atmospheric air.

5thly. Upon the direction of the wind. Sounds which are inaudible when the wind blows away from the hearer, are often distinctly heard when the wind blows towards him.

6thly. Upon the quantity of sound brought to the ear. If the sound which reaches the ear, at any instant, come from a large number of points, it will, other things being equal, be more intense than when it comes from a fewer number of points. Thus in the violin, the case, as well as the strings, is put into a state of vibration by the action of the bow, and in this way, the sound being originated at a large number of points, the sensation is proportionately intensified. The quantity of sound reaching the ear may also be increased by reflection ; a person speaking in a room is in general more distinctly heard than in the open air, the walls and ceiling of the room acting as reflectors of the sound.

227. **Pitch of musical sounds.**—That difference in the quality of notes which is remarked, when some are termed high and others low, is shewn by experiment to depend upon the *time* of the vibrations. The higher the note, the greater the number of vibrations per second. Various contrivances have been employed for the purpose of learning the number of vibrations corresponding to any given note ; of these, the two most important are Savart's apparatus and the Siren.

SAVART'S APPARATUS consists of a toothed wheel A, supported in a strong frame, and connected by an endless band B passing

round its axle, with another wheel C, so that when the wheel
C is set in motion by the handle D, the toothed wheel A can

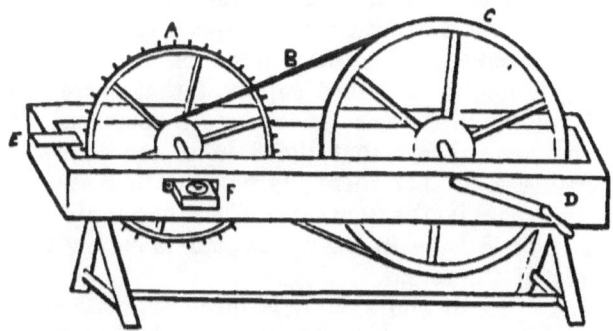

be made to revolve with great velocity. A piece of card, or a
thin elastic plate, is placed at E, in such a way that each tooth
in succession strikes against it, and causes it to vibrate. There
will, consequently, be as many vibrations of the card, during
one revolution of the wheel A, as there are teeth in the wheel.
An index at F records the number of times the wheel A has
revolved. In order, then, to learn the number of vibrations
corresponding to any particular note, we have but to put the
wheel into motion until the desired note is attained, and then
maintaining for awhile the same speed, we can observe by the
index the number of revolutions of the toothed wheel during a
second of time, and this, multiplied by the number of teeth,
will give the number of vibrations required.

THE SIREN consists of a small metal cylinder A, bounded by
two plane surfaces. The lower plate is pierced
by a short pipe D, through which a current of
air can be sent into the body of the instrument,
either from the mouth or by means of bellows.
The upper plate is pierced by a number of small
holes equally distant from each other, and from
the centre of the plate. Upon this plate rests
another plate C, pierced by a corresponding
number of holes, and moveable about a vertical
axis B. The holes in both plates are made as
shewn in the annexed figure, that is, not perpendicularly to the plates, but inclined in opposite
directions, so that when a current of air passes
through the lower hole, it strikes against the sides of the upper

one, and so puts into motion the plate C. Let us now suppose that the number of holes in each plate is
8. Each hole is successively opened and closed
8 times, during one revolution of the plate C,
and as each opening and closing gives rise to
a vibration, it might be hence supposed that there would be
8 times 8, or 64 vibrations during one revolution. It must,
however, be remembered that the 8 holes are all open or all
closed together, and that there are no more *different* openings
and closings than if the moveable plate had only one hole.
The effect of the 8 holes in the upper plate is merely to increase
the intensity of the sound.

The upper part of the axis B is furnished with an endless
screw, and this is connected with wheel-work contained in the
box or chamber E E, so contrived that two pointers in the front
shall indicate, the one the number of revolutions of C under a
hundred, and the other the number of hundreds of revolutions.
The total number of revolutions, as shewn by the pointers,
multiplied by 8, the number of holes, will give the total number
of vibrations during the period of observation. The inventor
of this elegant instrument was M. Cagniard de Latour, and the
name Siren was given to it because it emits sounds when placed
under water, and a current of that fluid is forced through it.

228. **Natural or diatonic scale.**—A succession of notes
whose vibrations are in the ratios denoted by the following
fractions :

$$1, \quad \tfrac{9}{8}, \quad \tfrac{5}{4}, \quad \tfrac{4}{3}, \quad \tfrac{3}{2}, \quad \tfrac{5}{3}, \quad \tfrac{15}{8} ; \quad 2, \quad \tfrac{9}{4}, \quad \&c.,$$

each successive seven being the double of the preceding seven,
forms what is termed the natural or diatonic scale. Each series
of seven notes in this scale is represented either by the letters
C, D, E, F, G, A, B, or by the syllables *do, re, mi, fa, sol,
la, si.*

It has been agreed that the note which is produced by 440
double vibrations, or 880 single vibrations per second, shall be
the note which is represented by the middle A of the treble
clef. Let A stand for this note, then the number of vibrations
corresponding to the seven notes of the scale are as follows :

|  | *do* | *re* | *mi* | *fa* | *sol* | *la* | *si* |
|---|---|---|---|---|---|---|---|
|  | C | D | E | F | G | A | B; |
| relative No. of vibrations— | 1 | $\tfrac{9}{8}$ | $\tfrac{5}{4}$ | $\tfrac{4}{3}$ | $\tfrac{3}{2}$ | $\tfrac{5}{3}$ | $\tfrac{15}{8}$ ; |
| absolute No. of vibrations— | 264 | 297 | 330 | 352 | 396 | 440 | 495 ; |

and if we represent the following seven by $C^1$, $D^1$, &c., we have

$$do^1 \quad re^1 \quad mi^1 \quad fa^1 \quad sol^1 \quad la^1 \quad si^1$$
$$C^1 \quad D^1 \quad E^1 \quad F^1 \quad G^1 \quad A^1 \quad B^1;$$

relative No. of vibrations— $\quad 2 \quad \frac{9}{4} \quad \frac{5}{2} \quad \frac{8}{3} \quad 3 \quad \frac{10}{3} \quad \frac{15}{4};$

absolute No. of vibrations— $\quad 528 \quad 594 \quad 660 \quad 704 \quad 792 \quad 880 \quad 990.$

**229. Octaves, fifths, fourths, &c.**—If starting with any note of the natural scale, we count onwards eight notes, the eighth is called the *octave* of the first. In like manner, if we count five notes, the last is called the *fifth* of the first; and so on for other similar terms.

Thus $C^1$ is the octave of C,   $D^1$ the octave of D;

   G   „   fifth   „   C,   A   „   fifth   „   D;

   F   „   fourth „   C,   G   „   fourth „   D.

It will be seen, that in every case, if one note be the octave of another, the number of vibrations of the two notes are in the ratio of 2 to 1.

It will also be seen, that in the majority of cases, if one note be the fifth of another, the number of vibrations of the two notes are in the ratio of 3 to 2. When this is the case, the notes are said to form a perfect fifth. Thus G : C, B : E, $C^1$ : F, $D^1$ : G, and $E^1$ : A are all perfect fifths; A : D, the ratio of whose vibrations are $\frac{5}{3} : \frac{9}{8} = \frac{3}{2} \cdot \frac{80}{81}$, differs so slightly from a perfect fifth, that it may be regarded as one. The remaining fifth $F^1$ : B, whose vibrations are as $\frac{8}{3} : \frac{15}{8} = \frac{3}{2} \cdot \frac{80}{81} \cdot \frac{24}{25}$, is called an imperfect fifth.

With a like exception, if one note be the fourth of another, the number of vibrations are in the ratio of 4 : 3. The exceptions are B : F = $\frac{15}{8} : \frac{5}{2} = \frac{4}{3} \cdot \frac{81}{80} \cdot \frac{24}{25}$, which is an imperfect fourth, and $D^1$ : A = $\frac{9}{4} : \frac{5}{3} = \frac{4}{3} \cdot \frac{81}{80}$, which does not sensibly differ from a perfect fourth. The ratios of seconds, thirds, sixths, and sevenths are less uniform.

N.B.—The ratio of the number of vibrations of one note to those of another is called in musical language the *interval* between the notes. An interval of $\frac{81}{80}$ is the smallest recognised in music, and is called a *comma*.

**230. Sharps and flats.**—If the number of the vibrations of a note is increased by 1–24th part (or multiplied by $\frac{25}{24}$), the note is said to be *sharpened;* if decreased by 1–25th part (or

N

multiplied by $\frac{2}{2}\frac{4}{4}$), it is said to be *flattened*. A sharp is denoted by the sign ♯, and a flat by the sign ♭. If the sharps and flats of each note were introduced, it would give rise to a scale of 21 notes. Some of these, however, differ so slightly that they are, for convenience, commonly regarded as identical. Thus C♯ and D♭, D♯ and E♭, E and F♭, E♯ and F, F♯ and G♭, G♯ and A♭, A♯ and B♭, B and C♭, B♯ and C¹, are treated as identical, and hence the scale is reduced to 12 notes; namely,

$$C \quad {C♯ \atop D♭} \quad D \quad {D♯ \atop E♭} \quad E \quad F \quad {F♯ \atop G♭} \quad G \quad {G♯ \atop A♭} \quad A \quad {A♯ \atop B♭} \quad B.$$

This scale of notes is called the *chromatic scale*.

**231. Sonorous waves.**—The length of a *wave* or *undulation* is the distance travelled by the sound in the time of a single vibration. Hence, the length of the wave corresponding to any note may be found by dividing the velocity of sound by the number of vibrations per second. Or if V be the velocity of sound, and N the number of vibrations,

$$\text{length of wave} = \frac{V}{N}.$$

Hence the higher the note the shorter the wave; and if the higher note be the octave of the lower, the length of its wave is exactly one-half that of the other; if the higher note be the fifth of the other, the length of its wave is exactly or nearly two-thirds that of the other; and in like manner of other intervals.

It follows also that the length of the wave of any note varies as the velocity of sound, and consequently is different in different media; is, for instance, greater in air than in carbonic acid; is nearly four times greater in hydrogen than in air; and somewhat greater in water than in hydrogen.

If we take the velocity of sound in air to be 1125 feet per second, the length of the wave of the note denoted above by C is 4·26 feet, that of the C next below this (or $C_1$) is 8·52 feet, that of the C next below this (or $C_2$) is 17·04 feet.*

**232. Sounds produced by the vibration of air in pipes.**—If the column of air contained in a pipe open at one

* In some books the student will find the length of these waves given at half the numbers stated above. This arises from the fact that by the length of a wave is then meant the distance travelled by sound in the time of a semi or single vibration; and not as above, of a complete or double vibration.

or both ends be put into a state of vibration by any means, the resulting note is found to vary with the length of the pipe, when the vibrations are excited by the passage of a current of air over the open end; they vary also with the velocity of that current.

That in all cases the sound arises from the vibration of the air, and not of the substance of the pipe, is shewn by the fact that, all other things remaining the same, the note produced is the same in pitch, whatever the material of which the pipe is composed.

The lowest note which can be sounded by any pipe is called the *fundamental* note of the pipe.

The fundamental note of any pipe open at both ends is the note whose semi-wave is equal to the length of the pipe.

The fundamental note of any pipe closed at one end is the note whose semi-wave is equal to *twice* the length of the pipe.

Hence, if two pipes of equal length be the one open and the other closed, the fundamental note of the open pipe is an octave higher than the fundamental note of the closed pipe.

Hence also, if the same pipe be sounded in different media, its fundamental note will be the higher in that medium in which sound travels with the greater velocity.

If the number of vibrations of the fundamental note be represented by $1$, then the other notes, which can be sounded by an open pipe, are those whose vibrations are represented by the numbers

$$1 \quad 2 \quad 3 \quad 4 \quad 5 \quad \&c. ;$$

but if the pipe be closed, they are those which are represented by the numbers

$$1 \quad 3 \quad 5 \quad 7 \quad \&c.$$

Thus, if C be the fundamental note of an open pipe, the notes which can be sounded by it are C, $C^1$, $G^1$, $C^2$, $E^2$, &c.

### 233. **Transverse vibrations of strings.**—In an important class of musical instruments, the sounds are produced by the vibrations of tightly stretched cords, which are made to vibrate transversely, or in a direction perpendicular to their length.

The fundamental laws of the vibrations of stretched cords are most conveniently illustrated by means of the *monochord*, which is the simplest possible of stringed instruments. It consists of a metallic cord, fastened at one end A, and kept stretched by a weight P suspended from the other end. The cord passes over two fixed bridges at B and C, and may be put into a state of vibration by the bow of a violin. D is a moveable bridge, by means of which the length of the vibrating cord can be regulated. MN is a box of thin wood, serving to increase the intensity of the sound.

First. Let a given weight be placed at P, and find by trial the length of cord required to produce a given note, say the note C, it will be found that to yield the note D the length must be reduced to 8-9ths of its former length; to yield the note E it must be reduced to 4-5ths; and so on, the length being inversely as the number of vibrations.

Secondly. Let the weight P be increased, the pitch of the note is found to be raised. If P be increased fourfold, the length of cord which formerly gave the note C will now give the note $C^1$, or the number of vibrations is twice as great as before. If for P we put a weight equal to P × $(\frac{3}{2})^2$, we obtain the note G, or the number of vibrations is $\frac{3}{2}$ times as great as before. If for P we put a weight equal to P × $(\frac{4}{3})^2$, we obtain the note F, or the number of vibrations is $\frac{4}{3}$ times as great as before. And so in like manner for other notes. Hence, generally, the number of vibrations is said to vary as the square root of the weight or tension in the cord.

Thirdly. Use another cord of the same material as before, but of a different diameter, then, the weight being unaltered, it is found that the thicker cord yields the lower note; if twice as thick, it yields an octave lower; if $\frac{3}{2}$ times as thick, a fifth lower; and so on, whence it is seen that the number of vibrations is inversely as the diameter.

Fourthly. If a cord of different material be used, of the same thickness, and stretched by the same weight as in the first case, it is found that for a given length the pitch is lowered, if the density of the cord be greater than before, but raised if it be less than before, and the result of experiment shews that the number of vibrations varies inversely as the square root of the density.

234. **Echoes.**—Sound, when it meets with any hard or elastic surface is reflected, and, as in the reflection of light, the angles of incidence and reflection are equal; hence it may happen, that a sound meeting with one or more of such surfaces is reflected back to the point whence it originated. Thus, let O be an origin of sound, and let ABC be portions of a spherical surface of which O is the centre; then all the sound which reaches A, B, and C from O is 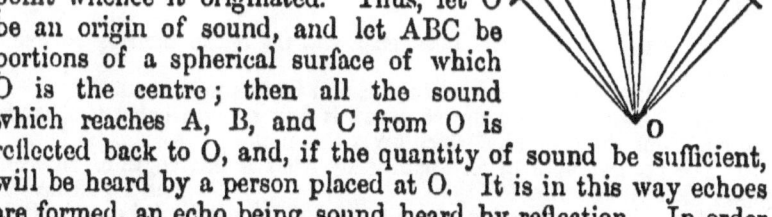 reflected back to O, and, if the quantity of sound be sufficient, will be heard by a person placed at O. It is in this way echoes are formed, an echo being sound heard by reflection. In order to the existence of an echo, three conditions are necessary :—

First. That a sufficient quantity of sound be reflected to the ear to produce sensation.

Secondly. That all the sound reflected from different points of the reflecting surface reaches the hearer at the same time. If part of the reflecting surface, C for instance, were more distant from O than A, the sound falling on both surfaces might still be reflected to O, but the sound from C would reach O later than that from A, and the result would be a confused, and not a distinct sensation.

Thirdly. That the interval between the origination of the sound and its return to the origin be sufficiently great to allow that the production of the sound should cease before any of the reflected sound reaches the hearer. If the space over which the sound travels in passing to and from the origin be considerable, time may be given for the utterance of a series of sounds before any of the reflected sound reaches the hearer; and, in such a case, several words may be uttered, and will be distinctly echoed. If the distance be short, there may be time for only a few syllables, or even for a single syllable. If the interval be so short as not to allow of the complete production of a sound before the reflection reaches the hearer, no distinct echo can of course be heard.

It is not necessary that the reflecting surfaces A, B, C be accurately spherical; for since a small portion of a spherical surface does not sensibly differ from a plane, a number of plane surfaces, situated at equal distances from O, may serve for the production of an echo.

## EXAMPLES.

1. A clap of thunder is heard 9 seconds after the lightning flash was seen, how far distant is the thunder cloud?          Ans. 1 m. 1615 yds.

2. If the velocity of sound in any medium be 1200 at a temperature of 70°, and 1165 at 60°, what is the velocity at the freezing-point?
Ans. 1067.

3. Shew that the interval between E and C$^1$♯ is the same as between D and B.

4. Shew that the interval between E and G♯ is a perfect third.

5. The lowest note in a church organ is four octaves below C, what is the length of the open pipe which gives this note?          Ans. 34·08 feet.

6. The highest note in an organ is B$^3$, what is the length of an open pipe which gives this note?          Ans. 1·7 inches.

7. If the length of a pipe whose fundamental note is E be 20 inches, what is the length of the pipe whose fundamental note is A?
Ans. 15 inches.

8. If the fundamental note of a pipe in air be C, what note will most nearly represent its fundamental note in hydrogen, assuming that the velocity in hydrogen is 3·7 times the velocity in air?          Ans. B$^1$.

9. If a pipe, whose fundamental note in air is A, has D for its fundamental note in a certain medium, what is the velocity of sound in that medium?          Ans. 27-40ths of the velocity in air.

10. Shew that the interval between D$^7$ and C$^1$ differs only from a perfect seventh by a comma.

11. If a stretched cord, 2 feet in length, yield the note D, what length of the same cord will yield the note A, the tension in both cases being the same?          Ans. 16⅔ inches.

12. If a cord when stretched with a tension of 18 oz. yield the note D, with what tension must it be stretched in order that it may yield the note G?          Ans. 32 oz.

# APPENDIX.

## A.—ADDITIONAL EXAMPLES ON CHAPTER I.

1. Resolve a force of magnitude 12 acting horizontally from left to right into two components, one of which is a force of magnitude 25 acting vertically upwards.—*U. of L. Matriculation, 1872.*

2. Shew that the resultant of the forces 7 and 14 acting at an angle of 120° is the same as the resultant of forces 7 and 7 acting at an angle of 60°.—*U. of L. Matriculation, 1872.*

3. The weight of a window-sash 3 feet wide is 5 lbs., each of the weights attached to the cords is 2 lbs.; if one of the cords be broken, find at what distance from the middle of the sash the hand must be placed to raise it with the least of effort.—*U. of L. First B. Sc. Exam., 1872.*

4. A body weighing 6 lbs. is placed upon a smooth plane, which is inclined at 30° to the horizon; find the two directions in which a force equal to the weight of the body may act to produce equilibrium. Also find what is the pressure on the plane in each case.—*U. of L. Matriculation, 1873.*

5. A heavy plummet is immersed in a stream, the string being held by a person on the bank. The string is found to settle in a sloping position. Shew by means of sketch the three forces which keep the plummet in equilibrium.—*U. of L. Matriculation, 1873.*

6. A particle is acted upon by a force whose magnitude is unknown, but whose direction makes an angle of 60° with the horizon. The horizontal component of the force is known to be 1·35. Determine the total force, and also its vertical component.—*U. of L. Matriculation, 1874.*

7. Shew that when five forces acting upon a point are capable of being represented in magnitude and direction by the sides of a pentagon taken in order they are in equilibrium.—*U. of L. Matriculation, 1876.*

8. Three cords are tied together at a point. One of these is pulled in a northerly direction with a force of 6 lbs., and another in an easterly direction with a force of 8 lbs., with what force must the third cord be pulled in order to keep the whole at rest?—*U. of L. Matriculution, 1876.*

## B.—ADDITIONAL EXAMPLES ON CHAPTER II.

1. A uniform rod 10 feet long is bent at right angles at a point 4 feet from one end. Find the perpendicular distances of the centre of gravity of the rod from the two straight portions of it.—*U. of L. Exam. for Women, 1872.*

2. Two uniform heavy rods, AC, BC, rigidly connected together, are capable of turning round a horizontal axis at C. Find the mechanical conditions which determine the position of equilibrium.—*U. of L. Matriculation, 1874.*

3. Weights of 2, 3, 2, 6, 9, 6 kilogrammes are placed in the angular points of a regular hexagon taken in order. Determine the position of their centre of gravity.—*U. of L. First B. Sc. Exam., 1874.*
Ans. Its distance from the centre of the hexagon is 5-14ths of one of the sides.

4. A rod AB weighing 10 lbs. is found to balance about a point 8 feet distant from A. A weight of 6 lbs. is fastened to A ; about what point will the rod now balance ?—*U. of L. Exam. for Women, 1874.*

5. Equal weights (each 1 oz.) are placed in the angular points of a heavy triangular lamina, and also at the middle points of the sides. Find the position of the centre of gravity of the plate and weights.—*U. of L. Matriculation, 1875.*

6. A short circular cylinder of wood has a hemispherical end. When placed with its curved end upon a smooth table it rests in any position in which it is placed. Determine the position of its centre of gravity.— *U. of L. Matriculation, 1875.*

7. A circular cylinder of uniform density, 4 feet in diameter, placed on a rough plane inclined at 30° to the horizon is just on the point of toppling over. What is its height ?—*U. of L. Exam. for Women, 1875.*

8. Squares are described on the three sides of an isosceles right-angled triangle. Determine the centre of gravity of the complete figure so formed.—*U. of L. Matriculation, 1876.*

9. A sphere of wood loaded at one point with lead rests upon a plane inclined at 30° to the horizon, being prevented from sliding down by the friction of the plane. State and explain by a diagram the conditions of equilibrium.—*U. of L. Matriculation, 1877.*

## C.—ADDITIONAL EXAMPLES ON CHAPTER III.

1. A wheel and axle is used to raise a bucket from a well. The radius of the wheel is 15 inches ; and while it makes 7 revolutions the bucket, which weighs 30 lbs., rises 5½ feet. Shew what is the smallest force that can be employed to turn the wheel. Upon what general principle is your answer founded ?—*U. of L. Matriculation, 1872.*

2. In the third system of pulleys in which each string is attached to the weight, each pulley weighs 3½ oz. Find the weight which will be supported by the pulleys alone when there are 5 movable pulleys.— *U. of L. First B. Sc. Exam. 1872.*

3. A uniform bar 20 inches long and weighing 2 lbs. is used as a common steelyard, the fulcrum being 5 inches from one end. Find the greatest weight which can be weighed with a movable weight of 4 lbs.—*U. of L. Exam. for Women, 1873.*          Ans. 14 lbs.

4. A substance is weighed from both arms of an unequal balance, and its apparent weights are 9 lbs. and 4 lbs. Find the ratio between the arms.—*U. of L. Matriculation, 1874.*          Ans. as 2 : 3.

5. In a system of pulleys in which a separate string passes round each pulley, it is found that when the power and the weight are in equilibrium,

and the power is caused to descend through 8 feet, the weight rises through 1 foot. Can the mechanical advantage of this arrangement be ever so much as 8? Give reasons for your answer.— *U. of L. Matriculation, 1875.*

6. A body whose mass is 5 kilogrammes, resting upon a smooth plane inclined at 30° to the horizon, is acted on by four forces—(1) its weight, (2) the reaction of the plane, (3) a force equal to the weight of two kilogrammes acting parallel to the plane and upwards, and (4) a force P acting at an angle of 30° to the plane. Determine the value of P, that the body may be in equilibrium.— *U. of L. Matriculation, 1877.*

This problem is most readily solved by resolving P into two forces — acting perpendicularly to the plane and along the plane. The later component is $\frac{1}{2}$ P$\sqrt{3}$, and the problem is reduced to the case of a weight 5 supported on the plane by a power $2 + \frac{1}{2}$ P$\sqrt{3}$, and these must be as the length of the plane to the height, that is, in this case as 2 : 1.


# D.—ADDITIONAL EXAMPLES ON CHAPTER IV.

1. The speed of a railway train increases uniformly for the first 3 minutes after starting, and during this time it travels 1 mile; what speed (in miles per hour) has it now gained, and what space did it describe in the first 2 minutes?— *U. of L. Matriculation, 1872.*
Ans. 40 miles per hour; $\frac{4}{9}$ mile.

2. In the last question, supposing the line level, and disregarding friction and the resistance of the air, compare the force exerted by the engine with the weight of the train.— *U. of L. Matriculation, 1872.*
Ans. The force is to the weight as 11 : 1080.

3. A rifle is pointed horizontally with its barrel 5 feet above a lake; when discharged the ball is found to strike the water 400 feet off. Find approximately the velocity of the ball.— *U. of L. Matriculation, 1873.*
Ans. 715$\frac{1}{2}$ feet.

4. Find the tension on a rope which draws a carriage of 8 tons weight up a smooth incline of 1 in 5, and causes an increase of velocity of 3 feet per second.— *U. of L. Matriculation, 1873.*　　Ans. 2 tons. 7 cwt.

5. A weight of 3 oz. hanging by a string over the edge of a smooth table draws another weight from rest along the table over a distance of 2 feet 6 inches in 5 seconds; find the weight on the table.— *U. of L. Examination for Women, 1873.*

6. In Attwood's machine one of the boxes is heavier than the other by half an ounce; what must be the load of each in order that the over-weighted box may fall through 1 foot during the first second?— *U. of L. Matriculation, 1874.*

7. The weights at the extremities of a string which passes over the pulley of an Attwood's machine are 500 and 502 grammes; the larger weight is allowed to descend, and 3 seconds after motion has begun 3 grammes are removed from the descending weight; what time will elapse before the weights are again at rest?— *U. of L. First B. Sc., 1874.*
Ans. 5$\frac{18}{19}$ sec.

8. A ball has a vertical velocity of 50 feet per second and a horizontal velocity of 70 feet per second simultaneously imparted to it; find the range and time of flight.— *U. of L. Examination for Women, 1874.*

9. A balloon is ascending vertically with a velocity which is increasing at the rate of 3 feet in a second ; find the apparent weight of 1 lb. weighed in the balloon by means of a spring balance.—*U. of L. Examination for Women, 1874.*

10. What is meant by saying, with reference to gravity, $g = 32$ ? What would be the value of $g$ if your units of space and time were miles and minutes ?—*U of L. Matriculation, 1875.*

11. A mass of 200 grammes is acted on by a force equal to the weight of 10 grammes for 20 seconds ; what distance will the mass have passed through, and what velocity will it have acquired (the acceleration due to gravity, 980 centimetres per second) ?—*U. of L. First B. Sc., 1875.*

12. A body projected vertically upwards against gravity has risen 120 feet in one second ; what was its initial velocity of projection, and how far will it rise during the next second ?—*U. of L. Matriculation, 1876.*

13. Two masses of 48 and 50 grammes respectively are attached to the string of an Attwood's machine, and, starting from rest, the larger mass passes through ten centimetres in one second ; determine from these data the value of the acceleration due to gravity, your units being centimetres and seconds.—*U. of L. Matriculation, 1876.*

14. A stone is thrown into the air at an angle of 45° to the horizon with a velocity of 128 feet per second ; shew that the path of the stone will not be a straight line, and determine the amount of vertical deviation from a straight line at the end of two seconds, neglecting the resistance of the air.—*U. of L. Matriculation, 1876.*

15. Supposing the unit of force to be that force which, acting upon 1 pound for 1 second, produces a velocity of 1 foot per second, state the relation between the unit of force and the weight of 1 pound.—*U. of L. Matriculation, 1877.*

16. A particle is projected in a horizontal direction with a velocity of 10 miles an hour, and at the same time falls under the action of gravity. Assuming that no other forces are acting, draw a picture representing the position of the particle at the end of 1, $1\frac{1}{2}$, $2\frac{1}{4}$, and 3 seconds.—*U. of L. Matriculation, 1877.*

## E.—MISCELLANEOUS EXAMPLES.

1. If two forces, 20 and 15, act upon a point at an angle of 60°, what is the magnitude of their resultant ?          Ans. 30·4.

2. If two forces, 10 and 12, act upon a point at an angle of 45°, what is the magnitude of their resultant ?      Ans. 20·34 nearly.

3. If two forces, 20 and 10, act upon a point at an angle of 30°, what is the magnitude of their resultant ?      Ans. 29·1 nearly.

4. If a bar, whose length is 20 inches and weight 16 ounces, be used as a lever of the first kind, find the distance of the fulcrum from the weight, when a power 4 ounces sustains a weight 60 ounces.    Ans. 3 inches.

5. Find the pressure upon the surface of a globe 12 inches in diameter, when immersed in water so as to be just covered.   Ans. 98 lbs. $3\frac{3}{4}$ oz.

6. A weight of 9 lbs. descends, drawing up another of 7 lbs. over a fixed pulley ; find the space described in 5 seconds.    Ans. 50 feet.

7. What horizontal force will sustain a weight of 240 lbs. on an inclined plane rising 5 in 13 ?           Ans 100 lbs.

8. Three parallel forces, 6, 8, and 10, act upon a body at distances 10, 12, and 14 respectively from a point without them ; what is the distance of their resultant from the same point ? Ans. 12½.

9. If the arms of a bent lever, of uniform thickness and density, be 8 and 10 inches respectively, and the distance between their bisections be 6 inches, find the distance of the centre of gravity of the lever from the bisection of the shorter arm. Ans. 3⅜ in.

10. If a cubic vessel, whose edge is 12 inches, be filled with a liquid whose specific gravity is 2·368, what is the pressure upon a side of the vessel ? Ans. 74 lbs.

11. If the focal length of a double convex lens is 5 inches, the radius of each surface being 8 inches, what is the refractive index of the material of the lens ? Ans. $\frac{9}{5}$.

12. The pressure upon a surface containing 54 square inches is 125 lbs.; what is the depth of its centre of gravity below the surface of the water ? Ans. 5 ft. 4 in.

13. If a cord 18 inches in length be fastened at two points in the same horizontal line, distant 12 inches apart, and if a smooth ring upon the cord sustain a weight of 20 lbs., what is the tension in the cord ? Ans. 13·4 lbs.

14. A mixture of gold and silver weighs 20 ounces, and its specific gravity is 12 ; what is the quantity of gold in the mixture ? Ans. 5·48 oz.

15. The pressure upon a surface containing 20 square inches, at a depth of 12 inches, is 15 lbs.; what is the specific gravity of the fluid ? Ans. 1·728.

16. Two convex lenses are placed 10 inches apart ; the focal length of the first is 3 inches, and that of the second 5 inches ; an object is placed 5 inches in front of the first lens ; find the position of the image formed by the second lens. Ans. 5 in. in front of the second lens.

17. If two equal bars, of uniform thickness and density, be used in combination as levers of the first kind, and if the length of each of the longer arms $= a$, and of each of the shorter arms $= b$, and the weight of each bar $= w$, shew that, when P and W both act downwards $Pa^2 + \frac{1}{2} w$ $(a^2 - b^2) = W b^2$ ; and when P acts downwards and W upwards, $Pa^2 + \frac{1}{2} w$ $(a - b)^2 = W b^2$ ; and when P acts upwards and W downwards, $Pa - \frac{1}{2} w$ $(a - b)^2 = W b^2$.

18. What must be the diameter of a screw, the distance between the threads being ½ inch, in order that the mechanical advantage may be 44 ? Ans. 7 in.

19. With what velocity must a body be projected vertically upwards that it may rise 225 feet ? Ans. 120 ft.

20. If a bar, whose length is $2l$ and weight W' be used as a lever of the first kind, find the distance of the fulcrum from the centre of the bar, when a power P sustains a weight W. Ans. $\dfrac{(W - P)l}{P + W + W'}$.

21. If three pulleys, whose weights are $w_1$, $w_2$, and $w_3$, be arranged according to the first system, and P and W be in equilibrium, shew that if P descend through a space $a$, and $x_1$, $x_2$, $x_3$, be the spaces through which the pulleys severally rise in consequence, then

$$Pa = Wx_1 + w_1 x_1 + w_2 x_2 + w_3 x_3.$$

22. A spherical balloon, 42 feet in diameter, is filled with a gas, whose specific gravity is ·9 ; what is the buoyancy of the balloon ?
                                                    Ans. 296·98 lbs.

23. A cylindrical body, floating in water, with its axis vertical, is immersed, when under an exhausted receiver, to three-fourths of its length ; through what proportion of its length will it rise when the air is admitted, supposing the specific gravity of air in comparison with water to be ·00122 ?                                    Ans. 61 – 199756ths.

24. If in the preceding, the cylinder, when under the receiver, be immersed to only one-fourth of the length, shew that the distance through which it will rise when the air is admitted will be three times as much as before.

25. A cylindrical vessel, 8 inches in height, and 12 inches in diameter, is filled with water ; the eye of an observer, who can just see the centre of the bottom of the vessel, is 30 inches from the rim, at what height is it above the level of the water ?                                    Ans. 18 inches.

26. If a ray of light, within a glass prism, whose section is a regular octagon, fall upon one of the sides at an angle of $45°$, and in a plane at right angles to the axis of the prism, shew that the ray can never emerge.

27. If ABC, forming an equilateral triangle, be the section of a triangular glass prism, and if a ray of light fall upon the side AB, in a direction parallel to the side BC, and at a point near to B, shew that the refracted ray will not emerge when it reaches the side BC, and also that it will emerge from the side AC, and in a direction parallel to that of the incident ray.

28. A bent lever, whose arms are 7 and 33, inclined to each other at an angle of $120°$, is in equilibrium under forces acting at right angles with the arms ; find the pressure upon the fulcrum when the power is 10 lbs.
                                                    Ans. 52$\frac{2}{3}$ lbs.

29. A heavy body is moving with an acceleration of 20 feet per second under the action of a force of 40 lbs.; find (i.) the weight of the body, and (ii.) the force required to give it an acceleration of 25 feet per second.                                    Ans. (i.) 64 lbs., and (ii.) 50 lbs.

30. Two cords, each 3 feet in length, are fastened at two points in the same horizontal line 4 feet apart ; the other extremities of the cords are attached to a weight of 10 lbs. : find the tensions in the cords.
                                                    Ans. 3√5 lbs.

31. A beam AB 20 feet long and weighing 10 lbs. rests on two props, one placed 2 feet from A, and the other 3 feet from B ; find the pressures on the props when a weight of 40 lbs. is suspended from A and a weight of 50 lbs. from B.                                    Ans. 40 lbs. on A, and 60 lbs. on B.

32. A bent bar ABC of uniform thickness and density having the arm AB double of the arm BC is suspended from A ; shew that when at rest the vertical line through A will cut BC at a point D such that BD – $\frac{1}{4}$BC.

33. A stone falling from rest is observed to pass over 112 feet in a second ; find (i.) how far it had previously fallen ; and (ii.) how far it would fall in the following 4 seconds.        Ans. (i.) 144 ft. ; (ii.) 768 ft.

34. If 101 litres of gas at $30°C$ be heated to $60°C$, what will be the increase of volume, the pressure remaining the same ?    Ans. 10 litres.

35. If 10 weight units of iron at $100°C$ be placed in 20 units of mercury at $50°$, what will be the temperature of the mixture, assuming the specific heat of iron to be ·1138, and of mercury ·033 ?        Ans. 81°·6C.

36. A closed vessel containing gas at 87°C is lowered in temperature down to 27°C, what is the diminution in the elastic force of the gas, supposing it originally to have been equal to the weight of 30 inches of mercury? Ans. 5 inches of mercury.

37. If a weight of 10 lbs. rest on a table, and the table descend with an acceleration of 16 feet per second, what is the pressure of the weight upon the table; also what would it be if the table ascend with the same acceleration? Ans. 5 lbs., 15 lbs.

38. What weight of cork (sp. gr. ·28) will just float in water 7 oz. of marble, sp. gr. 2·8? Ans. 1¾ oz.

39. If the receiver of an exhausting air-pump be 9 times as large as the barrel, what is the pressure of the air after 5 strokes if at the commencement it were 15 lbs. per inch? What also if the pump be condensing? Ans. 8·857 lbs., 23⅓ lbs.

40. If 1 lb. of steam at 100°C be mixed with 5 lbs. of ice at 0°, what will be the temperature of the product? Ans. 40½C.

41. An inch cube of a substance of specific gravity 1·2 is immersed in a vessel containing two fluids which do not mix (the specific gravities of these fluids are 1·0 and 1·5); find what will be the point at which the solid will rest.—*U. of L. Matriculation, 1876.*

42. Given a concave spherical mirror, how could you find its radius of curvature by optical means alone, and without resorting to geometrical operations?—*U. of L. Matriculation, 1876.*

43. A square plate, whose area is 64 square inches, is immersed in sea-water, its upper edge, which is horizontal, being 12 inches below the surface; determine the whole pressure of the water upon the plate when it is inclined at 45° to the horizon, assuming a cubic inch of sea-water to weigh 0·63 ounces.—*U. of L. Matriculation, 1876.*

44. A rod of uniform section is formed partly of platinum (sp. gr. = 21·0) and partly of iron (sp. gr. = 7·5), the platinum portion being 2 inches long; what will be the length of the iron portion when the whole floats in mercury (sp. gr. = 13·5) with the iron 1 inch above the surface?—*U. of L. Matriculation, 1876.*

45. 500 cubic centimetres of oxygen gas are measured when the temperature is 20°C, and the temperature is then raised to 40°C, the pressure meanwhile remaining invariable; what is the volume of the oxygen at the latter temperature (the coefficient of expansion is $\frac{1}{273}$)? —*U. of L. Matriculation, 1876.*

46. A pound of common salt and a pound of water, both at temperature 15°C, are mixed together and the salt dissolved; does any change of temperature take place? Give reasons for your answer.—*U. of L. Matriculation, 1876.*

47. Given the focal length of a convex lens, explain generally how it is possible to find the size of the image of the sun which such a lens will give. In what respect will this image be altered by diminishing the area of the lens without altering its curvature?—*U. of L. Matriculation, 1876.*

48. A ray of light passes from air into glass, the refractive index of glass with regard to air being 1·5. Given the angle of incidence at the common surface, draw a diagram to shew how the angle of refraction may be accurately determined.—*U. of L. Matriculation, 1876.*

49. A solid of which the volume is 1·6 cubic centimetres weighs 3·4 grammes in a fluid of specific gravity 0·85 ; find the specific gravity and weight of the substance.—*U. of L. First B. Sc., 1876.*

Ans. 3·975 and 4·76 grammes.

50. A small object 0·1 inch long is placed at the distance of 3 feet from a convex glass lens of 12 inches focal length ; what is the length of the image and the distance of this from the lens?—*U. of L. First B. Sc., 1876.*

51. A solid weighs 320 grammes in vacuo, 240 grammes in distilled water at 4°C, and 242 grammes in water at 100°C, of which the density is 0·959 ; find the volume of the solid at these two temperatures, and deduce therefrom its mean coefficient of cubical expansion for 1°C.—*U. of L. First B. Sc.*, 1876.

Ans. 80 and 81·33 cubic centimetres ; coefficient of expansion = $\frac{1}{5150}$.

52. A certain quantity of air at atmospheric pressure has a volume of 2 cubic feet, the temperature being 55°F ; what does the volume of the air become when the pressure is increased by one-twentieth, the temperature meanwhile remaining the same?—*U. of L. Exam. for Women,* 1876.

53. Describe the process of determining the specific gravity of a liquid by weighing a solid in it. A piece of glass rod weighs 35 grammes in air and 21 grammes in water ; what is its weight in alcohol of specific gravity 0·9?—*U. of L. Exam. for Women,* 1876.

54. The focal length of a plano-convex lens is 10 inches, a small candle flame is placed on the axis of the lens at a distance of 25 inches from it ; describe by means of a diagram the nature and position of the image of the flame formed by the lens.—*U. of L. Exam. for Women,* 1876.

55. A body weighing 300 grammes (sp. gr. = 5) has 100 grammes of another substance attached to it, and the joint weight of the two in water is 300 grammes ; find the specific gravity of the attached substance.—*U. of L. Matriculation,* 1877.                    Ans. 2·5.

56. A thousand cubic inches of air at a temperature of 30°C are cooled down to zero, and at the same time the external pressure upon the air is doubled ; what is its volume reduced to, the coefficient of expansion of air for 1°C being ·00366?—*U. of L. Matriculation,* 1877.

57. It is found as a result of experiment that 25 grammes of copper, at a temperature of 100°C, are just sufficient to melt 2·875 grammes of ice at 0°, so that water and copper are finally at 0° ; find from these data the specific heat of copper, taking the latent heat of water to be 80.—*U. of L. Matriculation,* 1877.

58. The chief focal length of a lens is 12 inches ; how far must I place a luminous object from the lens in order to obtain an image twice as large every way as the object?—*U. of L. Matriculation,* 1877.

59. An air bubble at the bottom of a pond 10 feet deep has a volume of ·00006 of a cubic inch ; find what its volume becomes when it just reaches the surface, the barometer standing at 30 inches, and mercury being 13·6 times as heavy as the water of the pond.—*U. of L. Matriculation,* 1877.

60. What is meant by the refractive index of a substance? Explain the fact that an aquarium-tank appears to be much shallower (from front to back) than it really is ; and point out in what way the difference between the apparent and true thickness is connected with the refractive index of the water in the tank.—*U. of L. Matriculation,* 1877.

## F.—THE FRENCH SYSTEM OF WEIGHTS AND MEASURES.

The unit of length termed a *metre* is one ten-millionth part of the quarter circumference of the earth measured from the equator to the pole.

Its tenth, hundredth, and thousand parts are termed respectively decimetre, centimetre, millimetre.

Its multiples of ten, hundred, and thousand are termed respectively decametre, hectometre, and kilometre.

1 metre = 39·37079 inches.
1 yard = ·91438347 metre.

The unit of capacity, termed a *litre*, is a cubic decimetre.

Its subdivisions and multiples are denoted by the same prefixes as those used in the case of the metre.

1 litre = 1·76172 pint.
1 pint = ·5676275 litre.

The unit of mass is the quantity of matter contained in a body having a unit of weight ; and the unit of weight, termed a gramme, is the weight of a cubic centimetre of distilled water at its greatest density.

Strictly, a gramme is the thousandth part of a certain standard weight preserved by the French authorities, but the difference between this and the gramme as defined above is so exceedingly small that it may ordinarily be disregarded.

The subdivisions and multiples of the gramme are denoted by the same prefixes as those used with the metre and the litre.

1 gramme = 15·432349 grains.
1 pound avoir. = ·453593 kilogramme.

N.B.—In the French system the standard for specific gravities is distilled water at its greatest density ; and hence the specific gravity of any substance is the weight in grammes of a cubic centimetre of the substance. Hence also if a body be weighed in water at its greatest density the loss in grammes gives the number of cubic centimetres in its volume.